Why Read

Demathtifying

Demystifying Mathematics?

The focus is on simple but complete explanation of the reasons behind each concept and operation across the whole of school mathematics: from percentages to calculus. Answering fundamental questions that are rarely even asked turn school mathematics from a useless and unloved rote to vital and pleasing training in reasoning. Greatly improved performance naturally follows.

To name just a few items of the hundreds covered:

- What is the method behind the way we count? How computers use it to work faster
- Why is 90° important? (Without the answer, delving into Trigonometry is pointless.)
- Why is algebra a blessing, not a nuisance?
- Why things like '-1 × -1 = +1' are easily deducible, not commandments
- What is 'Chaos'? (Crucial for dealing with uncertainty)
- Ideas versus deduction, arbitrary definitions versus reasons
- Why not to worry if tested positive for a rare disease even if the test is 99% accurate
- How every scary mathematical term can be re-worded into a self-explanatory one

You won't just benefit from this book; you'll enjoy it.

Demathtifying

Demystifying Mathematics

Ilan Samson

QED Books

ISBN 1 85853 217 5

Published by QED Books

an imprint of Richard Griffin (1820) Ltd

Pentagon Place
195B Berkhamsted Road
Chesham, HP5 3AP
United Kingdom

www.mathsite.co.uk

Printed by Lightning Source International

Publisher's Disclaimer

We were unable to delete the jokes

So be warned, reading this book may be fun.

Contents

In plain English	**in *other* words...**	*page*

User instructions

Nothing here is difficult to understand.

It is however necessary to concentrate and pay full attention to every detail. If any item is not immediately clear, dwell on it and refer backwards if necessary (reading *on* will not make it clearer).

Each chapter is about a separate topic and can be regarded individually.

However, as each chapter assumes some knowledge from previous ones, they are best read in order.

By all means keep within reach of children, parents, teachers…

Overdose is harmless.

Chapter 1

Cool It

1.1 Fear of maths - not your fault, not maths' fault

It is an extraordinary phenomenon: historians listen to lectures about music; musicians read about the human body; chemists discuss history and psychology. Yet most non scientists not only cannot and will not do maths, they flee from the subject with unconcealed fear.

The average person, if threatened with a knife, would regard the situation as serious but not totally hopeless: he might escape, or fend off the attacker, or pacify him. But consider the same person faced with a mortal danger he could only survive by giving the correct answer to the question: " How much is a tenth of one per cent divided by ten to the power of minus three?"

It would most likely be his last living experience, as he would not even *attempt* to survive. A pity, for the complexity of this task is less than that of understanding an average paragraph in a newspaper.

The same thing holds for all school maths. Anybody *could* understand it, and certainly need not be petrified by it.

What then is the cause of this paralysing fear?

On the one hand, there are totally avoidable causes, such as bad teaching and obscure terminology. On the other there are some innate characteristics of the subject. Let us consider the latter first.

In subjects such as literature, history or politics, it is relatively easy to conceal ignorance. You can waffle your way through to some middling mark just by reading a book review. In maths however, each step is either right or wrong. No intermediate states exist to see you through. Five out of ten in maths is as meaningless as being half pregnant.

Also, one can rarely achieve anything in maths by a single step. Each step is usually quite simple, but there lurks the danger of succumbing to the tedium of their multiplicity, or of confusing their order. And when that happens, it is like hanging from a chain: one missing or faulty link and the world around you shoots upwards.

True, the merciless requirements of accuracy and the multi-staged completeness may be intimidating. **The problem however is one of sustained attention rather than the difficulty of each element.** This can therefore be overcome simply by a little determination rather than any special gift.

1.2 The real root of the problem

Let us now turn to the serious but avoidable causes of the fear of maths.

These arise from teaching practices which take little if any account of what, in the child's mind, is involved with coming to grips with new, reasoning-based concepts.

The following are some of the fundamental problems in the way maths is taught.

The typical message that is conveyed to pupils is that there is a specified way of performing certain tasks, the reasons for which are at best only fully understood by the teacher. So often, the reasons for doing things are again, at best only partially hinted at. In other subjects this would merely result in some deficiencies in knowledge; in maths it actually causes damage. Unless clear and convincing reasons are given for each step, pointing out conceivable alternatives, and showing what is wrong with them, a disturbing unease quickly sets in. The damage this causes is so serious that it might be preferable not to teach maths at all:

- Children realise that, unlike the arts, maths is reasoning-based, and therefore lack of understanding does trouble them.
- Being unaware of the deficiencies in the teaching, they attribute their distress to their own inability. (This is reinforced by the apparent calm of others in the class who will not admit that they too are baffled.)
- A resultant faulty performance does at least confirm that a problem exists. A more serious situation could arise when children find their work marked as correct when they have still not understood the reasons for doing what they did. This undermines the very concept of understanding. A lasting mystification ('Mathtification') thus develops, leading to a lack of confidence that inhibits any hope of remedying the relationship with maths in the future.

1.3 What it is actually all for

Worse still, most pupils don't even know why they are made to do so much maths anyway, and are too inhibited to even ask. No time is found for teaching them how to deal with faulty services, treat burns and other pains, (or, more importantly perhaps - re-install Windows...), although these are all things they will undoubtedly have to cope with at some time. Yet although most students will go through their whole lives without having to solve a single equation, or, God forbid, having to know what the cosine of 60° is, thousands of school hours are devoted to maths. Why? No one asks.

There is a good reason, but it is evident from the way maths is typically treated in schools that little regard is given it. It is true that most people may get away with never having to use the fact that cosine 60 ° is ½. So why teach it? Because what the student does benefit from is understanding why a concept such as this (confusingly named) cosine is important, and why other similar things are not. It is not the mathematical results in themselves that most people need. What they need is to have been trained in the deductive abilities which happen to be those required for deriving these mathematical results.

To put it more generally: everyone needs to be trained in accurate, multi-staged, reasoned thinking. Maths just happens to be the most convenient vehicle for training this ability.

1.4 Some particularly silly obstacles

What, other than a cruel conspiracy against children, could have created the silly, intimidating, and unnecessarily obscure terminology of maths?

True, obscure terminology is often in use (not just in maths), but this is viable only for those who have already understood the concepts they refer to. And so should it be with maths. To refer to concepts by strange sounding, unfamiliar terms at the stage where the student is still struggling to understand the essence of the concepts raises a totally unnecessary obstacle. A child ought never to hear the words Logarithm, Sine, Root, Calculus, Algebra etc. before he understands and is familiar with the concepts they refer to. When the fire alarm sounds in the school, children are not sent to a xygophryllomn; they are directed to the fire exit.

In mathematics too, all these despairingly mystifying terms can be replaced by familiar words which describe what these things actually mean.

Had not the Latin word percent been used when introducing the concept of rating relative quantities as hundredths to English speaking children, percentages would not be viewed as 'one of the subjects taught in school, in which you pass or fail....'

(Fortunately, we are allowed to learn how to multiply by 3 without loading the syllabus with a whole subject called something like tri-mult...)

Noting a couple of further points will help towards a fresh approach to maths.

There is a widely held belief that one is either mathematically gifted or doomed. If maths is properly taught, you don't have to be either. 'Mathematically gifted' children are really ones who are able to understand it mainly by themselves, or could be taught the concepts at a much younger age than normal.

The sight and sound of mathematical notation usually provokes fear in those who have suffered years of bad teaching. Even so, the monster can be unmasked.

If we agree, for the sake of economy, to write texts in the most abbreviated (abb.) but still understandable way possible, we could cut out 'connecting words', drop vowels, and replace long words with short signs. For instance:

$$\text{C4, } \not{\text{A}} \text{ BLDG1} \;\Rightarrow\; \text{BLDG2 V3?}$$

Presented with this, and allowed a few specific questions, you have the choice of coming up with some reasonable reaction, or be thrown off a cliff. What do you do?

You probably will have a go. You will probably ask "What are 'C4', A, and 'V3'?" The answer might be that 'C' stands for children; '4' their age; 'A' - adults; 'V' for vehicle (1 - pram; 2- bicycle; 3- motorcycle; 4- car...). You might even want to check that BLDG is building and not bulldog. Armed with this information, you have a good chance of working out the meaning:

'Can four-year old children, unaccompanied by adults, go from the first building to the second building by motorcycle?'

(Note that the reason for coding the text in this form, especially when used repetitively, is not to obfuscate or to create riddles. It is a way writing a story using only 19 symbols instead of 100. This is a central method by which maths make the processing of information easier, or even possible.)

In contrast, were the reprieve dependent on the correct choice between "true or false" regarding an abbreviated text that looks more 'mathematical', such as

$$\frac{0.5 \times A_y}{\sum_f A} > \frac{\max_f A}{N_f}$$

most would end at the bottom of the cliff, having not even asked any of the permitted questions. Why?

This text too is nothing but shorthand. If only they had stared terror in the face and tackled the problem in the same way, asked the legitimate questions "what do >, A, N, $\sum$, f, and y stand for?", they would have been told that A = age; y = your; $\sum$ = sum; f = 'in your family'; N = number (of people), and that > is a nice short way of writing 'greater than'.

Given this information that ordinary mortal could have at least delayed the day of his doom by working out that what he must do is take half his age, divide it by the sum of the ages of everyone in the family, and see if the result is greater than the result of taking the greatest (max) age in his family and dividing it by the number of people in the family.

So, accomplishing this mathematical-looking task is really no harder than making sense of the first piece of shorthand. In both cases it involves little more than asking questions about unclear items. There is no need to be less bold if the items look 'mathematical'.

You are not alone in needing this revelation. Offer someone £50 for telling you what Mashik of 45 is, and he will ask "what is Mashik?". Then offer him £100 to tell you what tangent of 45 is, and he will turn down the challenge. He will not even ask what tangent is, because it clearly sounds like maths, and we don't mess with maths... In fact, Mashik means tangent in another language, but because it is not recognisably mathematical it is not intimidating, and invites the natural response to unclarity - a question.

Abbreviated notations that require 'spelling out' exist in other areas too. In music, for example, you have pretty things like:

which spelt out in words amounts to a whole story about when and how to play a single note (involving also a fair amount of arithmetic). Yet millions of non-mathematical musicians handle such notations painlessly.

1.5 Dissipating the mist

What stands in the way of knowing why $\sqrt{\log_{0.5}(\sin 30^0)}$ is 1 or -1 is not the difficulty of *understanding* the ingredients, but reluctance to ask what they mean.

Remember: When faced with mathematical stuff you don't understand, just *ASK* - the earth will not erupt. The answers may be a little longer than "first right, second left...", but not much more difficult. And in any case, in maths the answers *exist*, (unlike to so many other questions we always ask...).

Chapter 2

Hundredths / 'Divide By A Hundred'

2.1 Why not simply use language we know?

We start with a little nonsense. *Percentages* are not really worthy of '*MATHEMATICS*'. So let's get them out of the way.

Every time you hear per cent think hundredths or divide by a hundred.

Regard the percent sign % a misprint: some beneficial dyslexia should rearrange the balls to read, more suggestively, as /oo .

As it really means just that, it is rather silly that dividing by a hundred has become such an issue.

2.2 Beware of skipping hidden steps

There is however another point that needs to be made here, and rarely is. We now know that 'eight per cent of twenty five' starts with 'eight divided by a hundred'. But what arithmetic function is implied by the word 'of'? And why do we use that word?

If, having to feed eight guests, you go to buy twin packs of pastry the name of which only Hungarians can pronounce, how would you go about telling the baker what you want? You would just point at them and say "Give me four *of* these."

What you mean is: "give me that delicacy-pair four *times* over" i.e. by *of* you mean *times.*

Armed with these revelations about *per cent* and *of*, the exacting task of calculating 'eight per cent of twenty five' is reduced to the mere child's play of 'eight divided by a hundred, times twenty five', or with less scribbling: $\frac{8}{100} \times 25$

These operations can be done in any order. In this case it is easier to do the multiplication (8×25) before the division (by 100). Surely, is it not necessary to divulge to the reader what the answer is. (2).

Percentage also turns up in the form of: "how many per cent is a quarter?"

Here too, we need only replace *per cent* with *hundredths.* (meaning '(one) divided by a hundred'). Thus the mind-boggling task of calculating *percentage* is reduced to 'how many hundredths is a quarter?'. Similarly, 'how many percent is six out of forty?' translates to 'how many hundredths is 6/40?'. ('out of' meaning 'divided by')

In general, 'how many percent is a given *ratio*, or *fraction*?' translates to 'how many hundredths equal the given fraction?'

The most convenient way of finding the answer to this question is to remember that fractions can be expressed in another form, namely by using the decimal point (decimal fraction). This form is convenient, because the second digit behind the point states how many hundredths we have, and the first digit behind the point how many tenths (and a tenth is ten hundredths).

For instance, 0.25 means

2 tenths (= 20 hundredths) and 5 hundredths: in all 25 hundredths.

The question was 'how many hundredths?'- the answer: 25.

Thus all we need to remember is how to convert an ordinary fraction (that is, one number divided by another) into a decimal fraction. The answer is short: long division. 6/40 worked out by long division gives, yes, 0.15, which means 1 tenth + 5 hundredths = 15 hundredths. This is the reason, the whole reason, and nothing but the reason, why six out of (i.e. divided by) forty is 15 per cent.

So the whole 'difficult subject' of *percentages* turns out to be nothing more than dividing by a hundred or counting hundredths.

2.3 Per cent traps

Simple as this may be, there is something that even experienced 'percentagers' often get wrong due to lack of attention.

When you are given a percentage, but are not told of what (i.e. 'times what', remember?) watch out! A common (and expensive) example of this is the misleading trader's lament.

If you buy something for £10 and sell it making a profit of 50% (50/100 = ½), surely that means that your profit is £5 and that you sold it for £15. If you manage to get 100% profit, it must mean that you made £10 on the deal, and sold the object for £20.

The 'ever suffering' trader however, who buys something for £10 and sells it to you for £20 will moan that he "only makes a 50% profit..." How come? What *he* means (but doesn't tell you) is that the £10 profit is 50% of the £20 - the *selling* price, not of the £10 *purchase* price. By the same token, if he buys something for £1 and sells it for £5, he would declare his profit to be 80%:

The profit (5 -1 = 4) divided by the selling price (5), i.e. 4/5 = 0.8 (= 0.80) = 80%.

You and I would call it 400% (!), namely:

A £4 profit divided by his *outlay* of £1. (4/1 = 4 = 400hundredths = 400%). Not bad.

Per cent *of what* is also the key to a common puzzlement: for example, why does adding 20% and then taking off 20% not get you back to where you started? What the 20% were ***of*** was not the same thing in both steps.

Say you start with 100, and add 20%. That 20% must be 20% *of something.* This something is the 100 *to which you do the adding*, so you add 20% of 100 (=20), giving you 120. When you now *take off* 20%, the % is, as ever, *of something*, and again the something is that *from which you now take it off*: the 120. This 20% *of* 120, i.e.

$20\% \times 120$ means

$20/100 \times 120 = 24$. Take this 24 from 120 and you are left with 96.

To summarize:

Adding 20% to 100 gives 120. Then, taking away 20% ***(from this 120)***, leaves only 96.

Thus if you keep adding 20% and then taking off 20% you make a small fortune. (From a big one.)

Chapter 3

Selfmult, Num.Mults

This is about what the conspirators call *Bases, raising* and *Powers,* in order to conceal the meaning of things that look like

3^5 $\qquad$ N^{-2} $\qquad$ $M^{1/3}$

By the time you have read the next few pages you will wonder why things like this could ever have mystified you. The only mystery that will remain is how nonsensical words like *base, raise, power, root* and *logarithm* came to be involved. We deal here with nothing more than shorthand for very simple and common calculations. They were invented for convenience only. We introduce them here for three reasons: first, we want to make use of this convenience; second, it is a simple way to show how one makes up these useful shorthands and how one interprets them in special situations; third, to show how easy it can be, with the proper presentation, to understand concepts which so many consider to be beyond their grasp.

3.1 Where does all this come from?

When prehistoric man needed to work out how many spray cans were needed for, say, four famous mammoth graffiti, each requiring 2 (CFC free) cans, he wrote 2+2+2+2. The inventory for immortalising seven mamm's was still manageable: 2+2+2+2+2+2+2. But for the whole gallery of 129, it got too tedious.

So someone had the idea of inventing a shorter form - using the '2' and the '129', but now writing the '2' only once, while still expressing how many cans, in total, would be sprayed into posterity. Someone suggested '2 129', but that might have been mistaken for 'two thousand one hundred and twenty nine'. Another proposed '2 T 129', yet another '2 ⌚ 129' etc.. Nobody remembers why, but eventually all agreed on '2 x 129', standing for 129 such 2s added to each other.

In short, forms such as 12 x 7 = 84 did not fall from heaven. Nor did the 'picture' 12^7. What happened was that just as there had been a need to *add* a number repeatedly, more times than was convenient to write, so there arose a need (indeed a very common need) to *multiply* a number (by itself) repeatedly. To give an example:

You have made an investment which doubles your capital each year (do let me know where). So you begin with the amount of your investment, which to save typing, we will call 'INV'. At the end of the first year you have 2 x INV. This will have doubled again at the end of the second year to give 2 x 2 x INV. By the end of the third year you will have 2 x 2 x 2 x INV. By the end of the 12th year you are very rich because you have over 4000 times your original investment (rich, that is, unless *everybody* is doing as well, in which case inflation will be such that all you have is useless paper printed with large figures).

But even, or especially, the rich, dislike the inconvenience of writing long snakes of '2 x'. So a shorthand form had to be found to write such *2 multiplied by itself,* say 101 of these, of course using the '2' and '101', but again writing the '2' only once.

And again, it was necessary that the notation not be confused with other notations having other meanings. Someone may have proposed '1012' , another '2♣101'. The one who shouted loudest suggested '2^{101}'. So that was what was agreed, and for no other reason that he shouted loudest. (By the way, had they cared that centuries later even common calculators could not display this they would have chosen something else.)

Of course this form holds not only for a 2 if 101of which are self-multiplied repeatedly; it is the same for, say, 9, where 27 of which are self-multiplied repeatedly, i.e. 9^{27}. In the investment we talked of above, your loot at the end of 12 years is written as 2^{12} x INV.

And so now you belong, contrary to all (non)expectations, to that prestigious elite that knows why 2^4 is sixteen, why 10^3 is a thousand, and why 10^6 is a million. Don't rush it: you might need a good few seconds to verify these figures (either in you head or with a pencil, but do either).

Beware! Even the best of badly taught students will be caught out by the task: 'Prove that P^4= P x P x P x P'

At best they will say they don't know how to prove it; at worst they will try. It *cannot* be proven, any more than one can logically prove that the name of the river that flows through London is 'Thames', because P^4 was *simply and arbitrarily chosen* as the short way to write P x P x P x P, just as arbitrary as the choice of '4 x P' as shorthand for P+P+P+P.

3.2 Now we can confuse everybody, but do we really have to?

So we now have an agreed written notation, but what do we call it in speech, if we are to avoid saying 'four times four times four times...' etc.? To refer to 4^6, verbally, we want a phrase that contains '4' (once) and '6'.

Here we have two options. The first is that we (again, arbitrarily) adopt some word such as 'toe', 'grill', or 'boredom', or equally nonsensically as it happens, 'base', 'raise', and 'power'. The other is to use words that actually describe what 4^6 is about.

Surely it makes sense to use terms that make sense, at least until one is familiar with and happy about the concepts in question. Note that we call 4 x 6 'four *times* six' because of the usual meanings of the word *times* (as in 'how many *times* do I have to tell you to ignore what I say?').

In the following, we pursue a very unconventional course that looks and sounds unlike anything normally encountered in maths. It has, though, the equally unusual benefit of enabling students to figure out, all by themselves, the meaning of some basic concepts in maths which, to most people, have so far remained eternally bewildering.

Once the meanings have been digested, one can then revert to the conventional forms.

(Readers who are familiar, and at ease, with the conventional terms & forms of 'powers' might find the rest of this section 3.2 superfluous. It will, however, still come in useful in later sections. All readers should regard the following as a temporary - ***temporary*** - measure that serves a crucially useful purpose in which the conventional means are mostly known to fail).

To express 4^6 verbally then, one would expect something to the effect of

"(repeatedly) self-multiplying 4, the number of multipliers is 6".

Note that we are not counting here the number of repeated *multiplications* that take place. In 4^3 (4 x 4 x 4), for example, there are only *two* acts of multiplication. What we count is how many of the *multipliers* take part in this orgy of (repeated) self-multiplication: 3 in this example.

Let us then abbreviate:

"(repeated) self-multiplying 4"	into	"***selfmult*** 4, and the
"number of multipliers is 6"	into	"***num.mults*** 6"
So, 4^6 would then be called:		"***selfmult*** 4, ***num.mults*** 6"

(Note that ***selfmult*** describes both what the '4' *does*, namely, *self-multiply*, as well as what the '4' *is*, namely, the *self-multiplier*).

The words ***selfmult*** and ***num.mults*** themselves suggest what they mean. So, from now on, if you came across, say, "***selfmult*** 5, ***num.mults*** 3", you could go ahead and figure it out (125), just from the meaning of these terms, rather than at the mercy of memorized rules.

But no, "***selfmult*** 4, ***num.mults*** 6" makes its meaning too obvious... So the child haters decided on:

'*Raise* (the *Base*) 4 to the *Power* of 6' (which is about as is as 'illuminating' as '*fry* 4 to the *sadness* of 6'...).

The *4* here is called *base*; the *raising* is what you do with it and the *6* is called *power*. What this operation has to do with '*bases*' '*raising*' and '*powers*' God only knows. The general idea must have been: If we use terms like this, children

a. will usually not quite remember what they stand for.

b. will get confused, then worry, and finally despair, so that

c. when they grow up they will avoid maths at all costs.

That is why *we* scrap this nonsense in favour of our new, self-explanatory, terms. (Yes, temporarily, until this and all the following concepts have been well digested.)

And we have only started: The demonstration of the full effectiveness of our new terms is yet to come, when it will de-mathtify two far greater obscurities.

Summarizing:

'***selfmult***' : The (repeated) self-multiplier (as a noun), or (repeatedly) self-multiplying, (as a verb)

'***num.mults***': How many there are of this ***selfmult*** (taking part in the repeated self-multiplication)

In general, the form A^b will be referred to as '*self-multed* number'.

Selfmult, Num.Mults

A word about a word, or rather lack of it. True, 'number of multipliers' is rather cumbersome, but the fault lies with want of an important word in the otherwise rich English language. 'Depth of pool' can more neatly be expressed as 'pool depth', 'price of monkey' - 'monkey's price' etc., but tidying up 'number of multipliers' as 'multiplier(s) number' is ambiguous: the 'number' could also identify the multiplier's rank (i.e. first, second, etc.) or simply its value. 'Multiplier amount' is not discrete, 'Multiplier quantity' is not definite; The nearest would be 'multipliers count' (as in 'head count'), but 'count' is also a verb. As this is maths and not poetry, clarity must take precedence over euphony, so in the absence of a suitable word for 'multiplier ###' (unlike *Anzal* in German, *minyan* in Hebrew), we have no choice but our 'number of multipliers' (num.mults). At least we always know, unambiguously, what it means. In maths, we have to.

Note that "*selfmult* 4, *num.mults* 6" actually has fewer syllables than

'*raise* 4 to the *power* of 6'

If *Num.mults* is only 2, as in 5^2, we can just say '5 *self-mult(ed)*'

There is a pair of conventional terms which can be considered with a little *more* kindness:

L^2 is also called *L squared*, and L^3 is called *L cubed*. There is *some* reason (albeit a very indirect one) for the use of the words *square* and *cube*.

What does it mean that 'the length of a line is 14 *cm*'? It means that 14 units, each 1 *cm* long, i.e. -------- , will fit into this line.

What does it mean that 'the area of a surface is 24 *sq. cm*'? It means that 24 area units, each the shape of a square with sides 1 cm long, i.e.

will fit into this surface.

A three dimensional object having 'a volume of 60 cubic cm.' (cu.cm) means that 60 little cubes, each 1 cm long, 1 cm wide and 1 cm high, i.e.

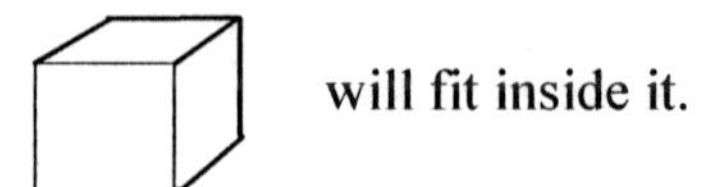

will fit inside it.

What, for instance, has an area of 24 *sq. cm.*? A rectangle 6 cm long and 4 cm wide. Why? Draw it and see: if its length were 6 cm and its width only 1 cm, the area would accommodate 6 *1cm by 1cm* squares (*sq. cm*); if 2 cm wide, another row of 6 *sq. cm.*; if 4 cm wide, a total of 6 x 4 *sq. cm.* But if instead of a rectangle we draw a square, remembering that the word *square* implies that all sides are of equal length, say 6cm, we have then 6 rows and the area of the square will be 6 x 6, or 6^2.

Because we happened to do LxL, i.e. L^2 to work out the area of a square with sides L, L^2 is called *L squared*. This is so because there happens to be something important to do with *squares*, their area, and that their areas are calculated by taking a number associated with the square (the length of its side) and multiplying it by itself. Were the salaries of cardinals calculated by taking their weight and multiplying it by itself, we might instead be calling L^2 'L cardinaled'...

Similarly, to work out how many *cubic cm* will fit into a box 4cm wide, 3cm deep and 5cm high, we first note that the area of the base is 4 x 3 *sq. cm*. Were this box only 1cm high it would contain 4 x 3 *cubic* cm. But being 5cm high, there are five such 1cm high layers, so the total volume, i.e. the number of *1 cubic cm* cubes that fit inside is 4 x 3 x 5.

If, however, the box were a cube, that is with width, depth and height all of equal, to 6 say, the cube's volume would be 6 x 6 x 6, or 6^3. Hence, by association with the method for calculating the volume of a cube, 6^3 is referred to as 6 *cubed.* Were the pope's salary determined by having his weight multiplied by itself and again multiplied by itself, we might be calling L^3 '*L poped'...*

3.3 Power games become child's play

We are now equipped to venture where, before, we did not dare to tread. In the past, faced with things like $7^3 \times 7^5$ or $A^k \times A^m$ we could at best look the other way and pretend it never happened. Now, having understood what selfmults (ex *raising* to *powers*) are about, we can laugh in the face of this monster.

3.4 Such obvious thought steps - why not use them?

Two simple rules should govern the undertaking of all such tasks, yet they are all too often ignored.

The first is to clarify the object of the task. It is common in schools to be set meaningless exercises such as 'do $7^3 \times 7^5$' or 'what is $A^k \times A^m$?'. The best answer to the latter is: 'it is $A^k \times A^m$!'. The best response to 'do $7^3 \times 7^5$' is: 'do what? colour it in? sing it? or what?'

The intended task in such exercises is usually 'write the following in a shorter, more concise way'. But whether or not this is the intended purpose of the task, never begin before you know what is actually required of you.

The second rule is to remind yourself of the *meaning* of every element in the given task. This is a simple and obvious rule, yet so many forget it, and then gaze helplessly at the task and think 'I have no idea what to do, and I never will know...'

What, then, do the components of $7^3 \times 7^5$ mean?

7^3 is 7 x 7 x 7 (three of these selfmultipliers), and 7^5 is 7 x 7 x 7 x 7 x 7 (five of them); so, $7^3 \times 7^5$ is 7 x 7 x 7 x 7 x 7 x 7 x 7 x 7; that is, how many 7s are there altogether repeatedly selfmultiplying? Total: 3+5=8.

Thus $7^3 \times 7^5 = 7^{3+5} = 7^8$. (That is to say, $7^3 \times 7^5$ is 'compacted' to 7^8.)

And what of $A^k \times A^m$? This means $\overbrace{A \times A \times A....}^{k\ A's} \times \overbrace{A \times A \times A....}^{m A's}$ We have, altogether, a 'chain' of k + m self-multiplying A's. How can we write this fact compactly? A^{k+m}

And: $P^7 \times P^q = P^{7+q}$, etc. Easy?

(*Note*: $7^3 + 7^5$, say, i.e. 7 x 7 x 7 + 7 x 7 x 7 x 7 x 7 *cannot* be compacted easily. Try).

Can 5^8 / 5^6 be compacted? Again, we disarm this monster by unmasking it, that is, by reminding ourselves what 5^8 and 5^6 mean:

5^8 is 5 x 5 x 5 x 5 x 5 x 5 x 5 x 5 $\qquad$ 5^6 is 5 x 5 x 5 x 5 x 5 x 5

$$\text{so, } \frac{5^8}{5^6} \text{ is } \frac{5\times5\times5\times5\times5\times5\times5\times5}{5\times5\times5\times5\times5\times5}$$

The order in which operations are carried out in such a sequence of multiplications and divisions does not matter. (Try it: for example, to work out

$$\frac{24\times10}{5\times6}$$

you get the same result (8) if you first do the multiplication 24 x 10 and then divide by 5 and again divide by 6, or if you first divide 24 by 6, then multiply by 10 and finally divide by 5, etc..)

If we do the above 'many 5' thing in the particular order of multiplying by a 5, then dividing by a 5, then again multiplying by 5, and again dividing by 5 etc. it would be as though we wrote it as

$$\frac{5}{5}\times\frac{5}{5}\times\frac{5}{5}\times\frac{5}{5}\times\frac{5}{5}\times\frac{5}{5}\times\frac{5}{5}\times5\times5$$

But each $\frac{5}{5}$ x is 1, so the above reduces to

1 x 1 x 1 x 1 x 1 x 1 x 5 x 5 i.e. 5 x 5

What has happened essentially is that each *dividing 5* cancelled the effect of one *multiplying 5*, leaving behind a (*doing nothing*) *1 times...*

In the example above, the six "*dividing by 5*"*s* turn six (out of the eight) *multiplying 5's* into 1 x... What is left is just two (eight less six) of the *multiplying 5* s. Thus

$$5^8 / 5^6 = 5^{8-6} = 5^2$$

Likewise C^p / C^f means

$$\frac{\overbrace{C\times C\times C\times C\times\ldots}^{p\ C\text{'s}}}{\underbrace{C\times C\times C\ldots}_{f\ C\text{'s}}},$$

every '$\frac{C}{C}$ x' is '1 x'

We assume for now that f is smaller than p. So f out of the p multiplying Cs turn into impotent little x 1s, and what remains is p-f of the multiplying Cs. i.e.

$$C^p / C^f = C^{p-f}$$

Any doubt about how useful this is? Imagine you need to work out 12^{106} / 12^{104} (by hand. Calculators are too lazy for this. Try.)

For a start, working out what 12^{106} and 12^{104} are would take care of the rest of your happy childhood, not to mention the remaining task, which would do justice to the term '*long division*'. But if you are late for a date, simply note that $12^{106}/12^{104}$ is $12^{106-104} = 12^2$ i.e. 144.

3.5 Every joke has a limit, but by rationalising we laugh on

Now for some exciting problems: From the definition of A^n we have

A^4 is A x A x A x A

A^3 is A x A x A

A^2 is A x A

That is, each time we reduce the ***num.mults*** on the left by 1, we drop an A from the chain A x A... on the right, until we are left with

A^1 (which) is (just) A.

But what is A^0 ???

Continuing the series downwards from $A^1 = A$, that is, going from *1* to *0* on the left hand side, and again dropping (the only remaining) A from the right hand side, what do we get? Blank paper...

And that does not have much meaning. In other words, we discover that the original 'story' about the form A^n does not help when n is 0. So we must take another route: We start by asking how A^0 could happen, that is, we ask how a ***num.mults*** of zero could come about. (It is not only in maths that the way out of a problem often is to ask 'how did I get *into* this mess?')

We *had* a situation in which the number of multipliers gets reduced: $A^8 / A^6 = A^2$, or in more general terms:

$$A^n / A^m = A^{n-m}.$$

Under what circumstances could n-m equal zero? The answer is: Only if *m* is equal to *n*.

In this case also the A^n and the A^m on the left are equal. The left hand side, then, is *something divided by itself*, which, even on a bad day, equals 1. It follows that A^0 must be 1!

This is nice: we started with something the normal interpretation of which is meaningless, namely: 'how many As multiplying each other? - None...' but then *did* end up with a meaning for it.

There is another way of looking at this: The normal interpretation failed for A^0. Another interpretation therefore had to be invented for this case. But we could not just choose anything, lest the implications conflict with other, already known truths. So we chose one that is consistent with the already established fact that A^n / A^m is the same as A^{n-m}.

And note that A^0 is 1 no matter what A is! In other words, also $4^0 = 1$; $197.5^0 = 1$ etc..

This brings us to another point. When do we use letters (or any other symbols) instead of numbers? We do so when we make a quantitative statement (such as '*certain two things are equal*'), and that the statement is correct irrespective of the numbers we apply to it. $Z^0 = 1$, for instance, is true no matter how much Z is. This principle is not confined to maths: to convey that a man, unaided, cannot cross England in one day, we would not say 'must use a bus'. A 'plane, a car, or, with a bit of luck, a train would serve him just as well. What we would say therefore is 'must use a vehicle', *vehicle*, being to *bus* what *Z* or *G* is to 5 or 0.712.

So whenever you are troubled by the sight of a Mathematical expression using letters, simply rewrite it, at least in your mind, substituting some *number* for the letter. This will help you digest the meaning.

More things can occur: how, for instance might we interpret W^{-3}?

The attempt would follow these lines: If *number of multipliers = zero* meant having no A's at all, then *number of multipliers -3* would mean that writing no A's at all is writing three too many. Which again is meaningless. So we start again: how could a ***num.mults*** of minus three occur? W^4/W^7, for instance, could do it, because it is equal to $W^{4-7}=W^{-3}$. But what does W^4/W^7 mean? It means

$$\frac{W\times W\times W\times W}{W\times W\times W\times W\times W\times W\times W},$$ where, after substituting

1 for each paired $\frac{W}{W}$ we are left with $\frac{1}{W\times W\times W}$

(The '1' is the same as the 1x1x1x1 left over from the *W/W*'s).

(Remember that there is no '*cancellation* of letters', it only looks as though there were. What happens is that *times W divided by W,* or W / W, leaves 1, or *1 times...)*

Once again we have managed to come up with a meaning for something which appeared not to have any: We found that W^{-3} means

$$\frac{1}{W\times W\times W}$$

We can of course write this more compactly: the *divider* $W\times W\times W$ can be written as W^3, so,

$$W^{-3}=\frac{1}{W^3}$$

You now belong to the super-breed which can meet what had seemed like horrific challenges, such as 2^{-1} and survive; not just survive, but also know that this is *a half*:

$$2^{-1}=\frac{1}{2^1}=\frac{1}{2}$$

Not only that: we possess the undreamed of powers of knowing that... 10^{-2} is *one per cent*!!!

(you remember of course that *per cent* is merely a subversively obscure way of saying *divided by a hundred*, or *hundredth*):

10^{-2} = $1/10^2$= 1/100 (i.e. *one hundredth*) = 1%

(Remember? Misreading the % symbol as $/_{00}$ is like removing the secret agent's sun-glasses...)

Note that $P^a + P^b$ e.g.: $P^4 + P^2$ $(= P\times P\times P\times P + P\times P)$

or $P^a - P^b$ e.g.: $P^4 - P^2$ $(= P\times P\times P\times P - P\times P)$

cannot be 'compacted' like $P^a \times P^b$ or P^a / P^b . Try.

(We used the term *divider* for the *U* in $\frac{S}{U}$; others, who prefer fancy to clarity, call it *denominator*...).

3.6 Turning dragon slaying into a turkey shoot, just by turning some words around

The, by now well digested, relationship enshrined in $8 = 2^3$, or to put it more generally, $A = B^c$ is also the source of further important concepts in maths. To show how we arrive at these, consider the following sentence:

"If Pedro shouts hallelujah a penguin comes"

This sentence establishes the connections between three elements: Pedro, *hallelujah* and the penguin. We can express the relationship in three ways, each time making a different one of the three the subject (as we would by re-pronouncing the above, each time changing the stressed word):

A *penguin* ***is*** what comes when Pedro shouts hallelujah.

Hallelujah ***is*** the call with which Pedro makes a penguin come.

Pedro ***is*** the lucky guy who gets penguins by shouting hallelujah.

Note that in each version, the *subject* is related to the other elements by ***is***. The mathematical shorthand for ***is*** is =.

We dealt above with a relationship (connection) between three numbers: 8, 2 and 3: $8 = 2^3$. This says: *8* (the *result*) ***is*** what we get if 2 is the self-multiplier and 3 is the number of these multipliers. The subject here is the *result* 8.

We can invert this 'sentence', just as we did with Pedro's penguin story. Let us start by making *2* (the self-multiplier) the subject:

2 (the *selfmult*) ***is*** what generates the result 8 if *num.mults* is 3.

In other words, *2* is the answer to the question 'what is the number, which, if three of it are self-multiplied, gives the result 8?'.

Or, 'what is the *self-multiplier* which, etc....'

Such questions keep coming up in maths. For instance, when 81 results from four equal multipliers multiplied together, what ***is*** the multiplier? The only way to find the answer is to try various numbers, multiply four of them by each other, and see if one gives the result 81. Try it. 1x1x1x1 doesn't get you very far from 1... 4x4x4x4 badly overshoots the mark. But 3? Bingo! 3x3x3x3 is 81.

We can shorten the wording for this question a little:

What is the self-multiplier for result 81, with num.of multipliers 4'?

Or, if on a really tight budget:

'***selfmult*** for 81 (***num.mults*** 4)'

(Remember: Whenever in any doubt about any of our abbreviations, just SPELL it OUT, and consider the meaning. That is the whole point of our new terminology: These terms *have* discernible meanings.)

Try one yourself: What is the ***selfmult*** for 32 (***num.mults*** 5)?

Again, $\overbrace{1\mathrm{x}1\mathrm{x}1\mathrm{x}1\mathrm{x}1}^{5}$ (i.e. ***selfmult*** 1) leaves you grounded; $\overbrace{3\mathrm{x}3\mathrm{x}3\mathrm{x}3\mathrm{x}3}^{5}$ goes through the roof, but... Ah, sorry, that one was yours.

Where ***num.mults*** is only 2, for instance '***selfmult*** for 36, with ***num.mults*** 2', one can just say '***selfmult*** for 36' (=6).

3.7 Uprooting mystifying language

Unfortunately, this concept, like others in maths, has undergone successful treatment in mystification of terminology. Nothing sufficiently obscure was found above ground, so they went digging and came up with *roots*. Our '***selfmult*** for 32, with ***num.mults*** 5' they called the '*Fifth* ***root*** of 32'... This is *bad.* There are no roots under 32, and you don't get 32 from 2 by planting 2 and watering it. You get 32 from 2 by letting 5 of the 2s engage in *self multiplication!* The 2 therefore is not a blossom, nor a root, it is a *self-multiplier* (***selfmult***).

(Note that the relevant vantage points for evaluating the relative merits of these terminologies, 'self-multiplier' versus 'root', are on hearing them for the first time, and on hearing them after having forgotten them.)

There is another way to root out this nonsensical term *root.*

There are two ways to arrive at 216 using three equal ingredients. One way is by taking 72 and *adding* three of them together (72+72+72 = 216). The 72 is called a *fraction* of 216. The other way to obtain 216 (again, from 3 equal numbers) is by taking *another* number, 6, and *multiplying* three of them together (6x6x6 = 216). Why not call this 6 a *fragment* of 216 ? Just as we use the term *fraction* for the equal components that are combined with each other by *addition*, so we use the term *fragment* for the equal components that are combined by *multiplication.*

(Clearly, *fragment* is then an alternative for our *self-multiplier.*)

To indicate how many of these *fragments* are needed - in this case three - we could say 'the 3-way *fragment* of...' So, the above example would be: 6 is the 3-way fragment of 216.

2 is the 4-way *fragment* of 16 (2x2x2x2, or 2^4). The equivalent form for *fractions* would be: '2 is the 8-fold *fraction*' of 16 (2+2+2+2+2+2+2+2, or 2x8).

There is another couple of terms for the parts of what something is composed of: *components* and *factors*. It is already a convention in maths to call parts *factors* if they are multiplied together. *Factors* however need not be equal to each other. So we suggest the use of the word *fragment* for *equal* factors.

(Note also the use of *way* in *'3-way'* to denote the *number* of *fragments. Way* (as in *which way?*) implies *direction.* We recall that to calculate volumes we need to multiply three numbers together, and these numbers are the lengths of the three dimensions, i.e. directions).

3.8. If it can *sound* right, it also need not *look* bad

Then there is the question of how to *write* this in a compact notation. As usual, an obscure and off-putting symbol, $\sqrt{A}$, was devised for the 2-way *fragment* of A.

A^2 implies finding the result of multiplying A by itself.

2-way Fragment of A implies finding what, if multiplied by itself, gives A. So, this is the 'opposite operation' of doing A^2. As such, surely, something like ^2A would make a much nicer and more natural notation than $\sqrt{A}$ (or $\sqrt[2]{A}$). However, we will use another form of 'oppositing' A^2 : namely, $A^{1/2}$, as this form actually *is* conventional, and this time with a good reason. For now we shall write this $A^{1/2}$ just 'for the looks', but shortly we will see why ***num.mults*** ½ *must* mean *2-way fragment* (ex. 'root').

So, let us bury roots. That is where they belong. $\sqrt{}$ might be used to catch fish, unless they too flee from it. *We* will write things like $32^{1/5}$ and call it the *5-way Fragment of 32.*

3.9 Plain sailing to feeling like masters of the universe

Finally, we are about to slay the biggest dragon of them all, a vicious word which has caused more widespread terror than Ghengis Khan: *Logarithm.*

Incredible as it may seem, given the choice of either defusing a bomb or answering the question 'what is the base2 logarithm of 8?', most people, even if they are allowed a few questions about the maths before answering, would opt for messing with the bomb. You will now see that this terror is due solely to the incomprehensible and frightening word *logarithm* used to refer to a perfectly simple concept.

Here it is: We started with $8 = 2^3$, which prescribes the arithmetic operation which connects the three numbers 8 , 2 and 3. We then changed the *subject* of this relationship from *8=...* to *2=...* ('*2 **is*** the 3-way *fragment* of 8', or, '*2 **is*** the *selfmult* for 8 with *num.mults* 3).

We can also, of course, make 3 the subject. This, quite simply, says:

'*3 **is*** the number of multipliers needed for result 8 if the self-multiplier is 2'.

Another way to state a *subject* is by way of a question, the answer to which is the subject. For instance: 'What is the number of multipliers needed for result 8 if the self-multiplier is 2 ?'. The answer is 3. Now *you* answer:

What is the number of multipliers needed for result 10,000 if the self-multiplier is 10?....

The answer, you will admit, was about as difficult as slowly counting to 4... (10x10x10x10 =...)

What you perhaps don't know is that you have just given the correct answer to the question 'what is the *base*10 *logarithm* of 10,000 ?'.

And what *neither* of us knows is why this term, which scares people more than bombs, must be used when plain English makes it so simple.

After a very little practise we may cope with some shortening to, say,

'What is the ***num.mults*** for 10,000 with ***selfmult*** 10 ?' to de-mathtify

'What is the ***logarithm*** of 10,000 with ***base*** 10 ?'

We started off with the (written) notation R= A^n ('R'- for 'result'), and went on to devise a verbal phrase to call it by, namely: R= ***selfmult*** A ***num.mults*** n.

We then inverted this phrase, first to A= ***selfmult*** <u>for</u> R (***num.mults*** n) and suggested for this a nicer notation, namely: A= $R^{1/n}$ (to replace the ghastly $\sqrt[n]{R}$),

and to the second inversion n = ***num.mults*** for R (***selfmult*** A). What is left is to devise a (short) notation for this one too. Here is a suggestion (as an exercise, showing how notations *should* be developed):

In maths, brackets are commonly used to mean *for* or *of.* e.g.: *tan of 45* is written as *tan(45),* so, instead of

'***num.mults*** for 1000' we just write ***num.mults(1000).***

And where do we put the *self-multiplier*? We really should write it to the bottom left of ***num.mults***, because that is also its relative position in A^n. So, the notation should be:

$_{10}$***num.mults*** (1000)= 3 $_{2}$***num.mults*** (32)=5

However, as undue respect for the conventional form of log_2 (32) = 5, we will shift the $_2$ (the *selfmult,* remember?) to the bottom *right*: i.e.:

num.mults $_2$ (32) = 5

Surely, this reminds us of the meaning a little better than the riddle above it...

Thus to summarise the three 'inversions' of the connection between 8, 2 and 3, we have:

Phrase	*Notation*	*Ex Phrase*	*Ex notation*
selfmult 2 ***num.mults*** 3 = 8	$2^3 = 8$	***raising*** 2 to the ***power*** of 3 =8	no change
selfmult for 8 (***num.mults*** 3) = 2 or 3-way ***fragment*** of 8	$8^{1/3}= 2$	3rd ***root*** of 8 =2	$\sqrt[3]{8}$ =2
num.mults for 8 (***selfmult*** 2) = 3	$n.mults_2(8)$ = 3	***base*** 2 ***logarithm*** of 8 =3	log_2 (8)=3

The big advantage of these new phrases is that their *contents enable you to work them out!* (Even if it takes some time). In contrast, no amount of 'thinking' will help to work out the '*ex*' phrases for those who forgot their meaning, never knew it, or never understood it.

And note: our *two* new terms ***num.mults*** and ***selfmult*** replace *four* old ones: **power, logarithm, base** and **root.** The above reveals that **power** is to **log** what donkey is to ass. Wouldn't it be nice if they had told you. And the same with **base** & **root** (both of which mean the same as our ***selfmult,*** just as both **power** and **logarithm** mean the same as our ***num.mults).***

Now, once familiar and at ease with the above *concepts,* one can *then* revert to the conventional **powers, logarithms, bases & roots**, or Buxtehude, if one fancies.

3.10 Mopping up the rest

Let us return again to the various mishaps that can befall the *number of multipliers*.

We know what the notation M^c stands for, and that the story has meaning only if *c* is a whole number, greater than 0. Since the story was meaningless when *c* was 0, we set out to find other grounds for determining what M^0 must be equal to. We then went on to do the same in the case where *c* had the audacity of being a *negative* number, e.g. M^{-3}.

But then there is no end to the perils of life...:

Earlier we used a notation looking like ***num.mults*** as fractions – we used $A^{1/2}$ just to 'look' like the 'opposite' of A^2 . But what happens if we are told that ***num.mults*** actually *is* a fraction? Does $M^{1/2}$, $M^{1/4}$ actually *has* a meaning? What would this meaning have to be?

When asking how many multipliers take part in a repeated self-multiplication, the answer 'not quite one' is not the sort of language serious people appreciate... So let us try again to find a different interpretation for the notation M^c in the calamitous event of *c* being a fraction. Here it is not easy to see how something like $M^{1/2}$ could have come about, which is the method we used to tackle M^{-3}. Instead, let us see if there is something we can *do* with $M^{1/2}$, something that *does* end up with a meaningful outcome.

Since the problem stems from the ***num.mults*** being a fraction, we may ask whether there is any way me might 'build it up' i.e. increase it (for instance by adding to it), in such a way that we end up with a *whole* number. Have we not already had a situation in which addition was done with ***num.mults***? Of course; $W^n \times W^m = W^{n+m}$.

(If *remember* this you do, but forget *why* this is so, do not attempt to use it; go back and find out its reason! The real purpose of this whole exercise is to convince you that you *can* understand where all these things come from.)

If, in $W^n \times W^m = W^{n+m}$ the *n* is ½ , and our aim is that the *resulting* ***num.mults*** (n + m) be a whole number, 1 for instance, then *m* must of course also be ½., i.e.

$$W^{1/2} \times W^{1/2} = W^{1/2+1/2} = W^1 = W$$

What this means is that $W^{1/2}$ is something which, if two of them are multiplied by themselves, give ***W*** itself. If ***W***, for instance, were 9, then $W^{1/2}$ would be 3, because 3x3 is ...., and if W is 25, then $W^{1/2} = 5$.

What about $W^{1/3}$?

Try multiplying *three* of them (by themselves) repeatedly:

$$W^{1/3} \times W^{1/3} \times W^{1/3} = W^{1/3+1/3+1/3} = W^1 = W$$

$W^{1/3}$ then, is 'that which if you (repeatedly) self-multiply three of them, gives W'.

Let us say, for example, that W is 1000. How much is $1000^{1/3}$? We know that

$$1000^{1/3} \times 1000^{1/3} \times 1000^{1/3} \quad (= 1000^{1/3+1/3+1/3} = 1000^1) \quad = 1000.$$

That is, if we multiply three of these $1000^{1/3}$ by themselves we get 1000. But what *is* this number which yields 1000 when three of it are multiplied together? Only 10 fits the bill.

That is the reason $1000^{1/3}$ must be 10.

Doesn't all this sound familiar? Heart-warmingly familiar? Of course: 10 is our *3-way fragment* of 1000! And now we see why our nice notation for this $1000^{1/3}$ (ex $\sqrt[3]{1000}$) does, as promised earlier, have a good reason.

Similarly: $25^{1/2}$ is the *2-way fragment of 25,* (ex $\sqrt{25}$).

Generalizing: $A^{1/n}$ can relieve us from the sight of the (ex) $\sqrt[n]{A}$.

Before we leave the topic of *selfmults* and *num.mults*, let us look at another couple of questions, with a view to learning *methods* of tackling problems.

What is $(A^3)^4$?

Always begin by identifying the source of the problem. Here the problem is having *two num.mults*. Next, let us ask: what *do* we know?

We do know how to deal with a single *num.mults*, so let us do just that. How? By temporarily concealing the other: We leave the ^{4} (on the outside), and instead of (A^3) we write '$'. This way, our $(A^3)^4$ becomes (temporarily) $\$^4$ which we confidently know to mean $\$ \times \$ \times \$ \times \$$.

Now we can safely put the (A^3) back in place of the $, giving us $A^3 \times A^3 \times A^3 \times A^3$. We also know that each A^3 is AxAxA. How many As are being self-multiplied in all? 3 x 4. Thus $(A^3)^4 = A^{3 \times 4}$ $(= A^{12})$. Easy?

Not only do we know that $A^n \times A^m = A^{n+m}$.

We now also know that $(A^n)^m$ m $= A^{n \times m}$.

3.11 Yes, *of course* this has useful applications

The main purpose of this study, as proclaimed at the outset, is to train in *accurate, multi-staged, reasoned thinking*, and to discover that it doesn't kill. But the above concepts such as ***num.mults*** (ex *logarithms*) for example, *do* also have useful applications. Therefore, fussing around with them so much without telling you what some of their uses are would be bad teaching. (Not asking would be bad learning.)

In the olden days, before the advent of free calculators coming with every other promotion, the tedium of long multiplication was the order of the day. How could life be made easier?

It would be nice, for instance, if multiplication could be replaced by *addition.* But how (using only what you already know) could this be done?

Have we not already encountered a situation in which the multiplication of certain items could be rewritten in a form that involved addition? Of course... $R^g \times R^f = R^{g+f}$!

How can this be employed for a rare event that would justify doing anything more demanding than estimating the total cost of two bus rides at £1 each: say, checking up on your broker who sold you 9175 shares at £13.325 each ?

First we need to note what is special in the case above in which multiplication could be traded for addition: both the multiplied items are, themselves, self-multiplications - of the *same* self-multiplier.

So, you can chose a number - *R* - as a self-multiplier (***selfmult***), then imagine an oracle that tells you what the ***num.mults*** is (calling it *c1,* say*)* so that R^{c1} will equal 9175, i.e. the oracle providing the ***num.mults*** *for 9175 with* ***selfmult*** *R.*

The good oracle then also tells you what, *for the same R*, the ***num.mults*** for 13.325 is, (calling it c2). Thus $R^{c2} = 13.325$. From here on it is plain sailing:

$$9175 \times 13.325 \text{ is}$$

$$R^{c1} \times R^{c2} \text{ which, as we expertly know, is } R^{c1+c2}$$

Now all that is left to do is the *addition* c1 + c2. Let us call the result c3, (so, $R^{c1} \times R^{c2} = R^{c3}$)

Before we discharge the omniscient oracle, we must still get it to tell us the value of ***selfmult*** *R* ***num.mults*** *c3* (i.e. the above R^{c3}) - the result of our multiplication.

The 'oracle' is in fact tables: long tables listing the *num.mults* of a given *selfmult* (usually 10) for all numbers. For 100 it is easy: the *num.mults* is 2; for 1000: the *num.mults* is 3 (any doubt?), but how does one work out the *num.mults* for, say, 8 (if the *selfmult* is 10)? That is to say: what is the *c* in $10^c = 8$? Not to mention the *num.mults* for 13.325?

This one, I fear, *is* a long story, and the list of prerequisites for understanding it is even longer - too long for this book. Such omissions are sometimes unavoidable, but what *must* be done is clearly declare the omission, and not let you think that you are missing the link because you are too dumb to understand it...(a common sin in teaching).

Long multiplication is long, but long division is much longer. No matter. ***Num.mults*** work here in almost exactly the same way: We now replace

$$9175 / 13.325 \text{ by } R^g / R^f.$$

No one knows better than we that this equals

$$R^{g-f}$$

i.e. this time *subtraction* rather than addition delivers us from difficulty.

Another important use of *num.mults* (ex logs) occurs in the following.

In many systems of cause and effect, when the magnitude of the cause changes in equal steps, so too does the effect commonly change by equal amounts. For example, if a long spring is extended by 2cm when pulled by 1 kg, every additional 1 kg pull will extend the spring by a further 2cm. Such a dependence, or relationship, is called *proportional*, or *linear* - *linear* because if one drew a graph of the *length versus pull*, i.e. for each stage of stretch we mark a point on graph paper in such a way that the distance the of point from the left-hand margin represents the pull (say one square for each kg) and the height of the point above the bottom margin represents the corresponding length of the spring (say one square for each cm. in length), the line joining these points is straight, or *linear*.

There are however dependencies where the 'payoff comes harder and harder'.

Feeding the ear, for instance. The human ear is designed to protect the species from certain extinction in discos. For each *perceived* equal increase in loudness, the actual intensity must be raised ten-fold! For the perceived loudness to increase by two equal steps, the intensity must be raised a hundred-fold. Three steps to the ear - thousand-fold output from the damned electric guitars.

If we want to draw a straight-line graph, this time for the relationship between the listener's suffering and the rock band's output, we must convert the intensity values from the ten-fold increases that are required for every equal, one step increase as perceived by the ear, to something that also grows in equal-sized step. That is, we need to change the progression of 10, 100, 1000, 10000 etc. to something like 1, 2, 3, 4... What might effect this conversion? Easy: for the graph, we use not the *intensity* values, but their ***num.mults*** with ***selfmult*** 10!

> ***Num.mults*** for 10 (***selfmult*** 10) is 1
>
> ***Num.mults*** for 100 (***selfmult*** 10) is 2
>
> ***Num.mults*** for 1000 (***selfmult*** 10) is 3

etc. So the perceived loudness is not proportional to the intensity, but to the *num.mults* for the intensity. Such a proportionality relationship is therefore said to be on a '***num.mults*** scale' (*logarithmic scale* in the old, wishfully forgotten, jargon)

3.12 Foregone ignorance that might have cost you dear

Finally, let us look at an interesting example of the ***might*** of *num.mults*.

How does an economy go mad? How does inflation of a million per cent come about? (1,000,000% = 10,000, so a loaf of bread that costs £1 at the start of the year costs £10,001 at the end of it). There are all kinds of economic theories about this, but can any physical cause really get 10,000 times worse?

There is a simple explanation, and a warning: once we start to accept price rises every time we go shopping all hell may break loose. Wild inflation may begin with price increases that are so (deceptively) slight that they do not alarm us, or awaken us to their potentially devastating effect.

If the shopkeeper raises prices by 8%, making an item which costs £1 rise to £1.08, a derisory 8p more, but does so every three days (and gets away with it because *we* have come to accept frequent rises), within a year we have that million % inflation. This is how it works:

The first rise is from 1 to 1.08. Three days later it rises again by the same 1.08 ratio to 1.166 (1.08 x 1.08 [$=1.08^2$] = 1.166) (of course, rounded *up* to 1.17).

You may well wonder why the rise continues by a fixed *ratio,* (in this case of 1.08), and not by a fixed *difference* of 0.08.

The answer is that our *perception of severity* is being exploited. When the price reaches 10, what further rise would *feel* like the (original) rise from 1 to 1.08? Not from 10 to 10.08 (i.e. an *addition* of 0.08), but from 10 to 10.8, i.e. the present 10 *multiplied* by 1.08 as before. When it reaches 100 (it soon *does*), the further, equally perceived, rise would now be from 100 to 108 - again, the same *ratio* (of 1.08), not the same *difference*. How we *feel* about these rises is what this is all about, and is what is exploited to the limit! (To smooth this madness along, when the pound reaches the worth of a penny they rename the currency, to help you not notice what is going on....)

At the third 3-day rise we reach 1.08 x 1.08 x 1.08 (1.08^3). And so on. 'Every three days' occurs more than 120 times a year; 1.08^{120} is more than 10,000! - one million %. That is how.

Furthermore, high frequency rises make inflation terribly sensitive to the effect of 'small inaccuracies'. A rise of 1.08 '*casually rounded off*' to 1.10 makes things *ten times worse still:*

1.10^{120} is nearly 10 times greater than 1.08^{120}. By way of comparison, *doubling* the prices every five weeks (10 times a year), will leave you *ten times* better off than the 8% rise every 3 days: 2^{10} (which is 1024) is ten times smaller than the 1.08^{120}.

So now we need no longer fear *num.mults* (ex. *powers / logs*). Just easy on the big ones.

Chapter 4

Writing Numbers

4.1 They should have told you this when you were four, so it is high time you find out

A brief foreword

This book is about explaining the reasons behind what we do in maths. They are mainly provided with the material that is being taught. Here, however, we deal with reasons by themselves because they relate to something that everyone already does know how to do – writing numbers, and yet, few know why they do it the way they do. Does it matter? Yes, because this probably is the trend setting, first experience in receiving maths instructions without their reasons. From the start it also conditions us not to seek them.

Also, while understanding the reasons will not get us any better at writing numbers, we will be able to expand our knowledge into some crucial areas, for example, how computers do it.

Note: nothing in the following is complicated, but the detailed spelling out requires progressively slow, patient attention as the story proceeds. Towards the end of this section we start encountering mathematical shorthands. It looks intimidating at first, but as it only abbreviates what already is understood appreciation of the elegance of this brevity takes over.

Why is the quantity :. written 3 and not 2 or, say, △ ??

If 3 stands for the quantity :. and 6 stands for ::: and 5 stands for ::. why is the number of days in the year written 365 i.e. why does the 3 in 365 stand for 3-hundred, the 6 stand for 6-ty and, altogether, what do we think we are doing by sticking such 3, 6, 5 to each other?

High time to find out - about this - and about other things of which most of us know more that...s than why...s.

Most of us are not even aware that we don't know why we use numbers in the way we do. Were there no digits 5, 6, 7, 8, 9 i.e. if we only had the digits 0, 1, 2, 3, 4, how would we count beyond 4? How would you write the number of fingers on one hand? And on both? If, like most, you cannot answer this correctly, you do not know why you use numbers the way you always have.

To do maths, or not to do maths - this might be a question for some, but all of us, surely, should at least understand why we use numbers in the way we do? So let us do just that.

An obvious way of writing the number of days in a week would be :::. and this is probably the kind of way it was done originally. It became apparent however that memorising a single symbol was easier than counting a lot of dots, so someone invented the graphic '7' to stand for :::. at the same time he invented the graphic, or 'picture''2' for : . There was no reason for these choices; the two symbols might just as well have been interchanged, and had they been, today you could write that you have 7 legs and nobody would laugh. The inventor of these symbols also had to invent names for them; he called '7' 'seven' but it might just as well have been 'neves' or 'mutzi'.

Since single symbols are so convenient, why then do we write the quantity called a dozen as '12' instead of as some single symbol?

Perhaps because by the time our 'Numberist' had invented ten pictures (0, 1, 2, 3, 4, 5, 6, 7, 8, 9) his twelve-year university grant had run out (at the academic productivity rate of one picture a year, and allowing for two sabbaticals).

More likely though, the reason is that a great number of symbols would be difficult to memorise, so it was decided to limit them. To the number of fingers.

Note: whatever 'pictures' are used, the first one must stand for zero because any counting system must include the state of nothing. (There are many academic reasons for this, but also a very practical one: we do all this for counting, and the thing we count most often is money, and the most common amount of it is 'none'...). So it is only the second 'picture' which stands for 'one', and therefore, when there are ten 'pictures', the last one stands for 'nine', i.e. one less than the number of pictures!

But let us imagine that there were only eight pictures, i.e. 0, 1, 2, 3, 4, 5, 6, 7 (remember: eight pictures does not reach the 'picture' 8 because the 0 is included in the count). How would we now count beyond 7, given no further pictures? We are free thinkers, so let us try to find a way:

Maybe we could proceed by repeating the pictures, but in reverse order, adding a marker ↓, e.g. 7↓, 6↓, etc. to distinguish them from their first appearance. This does keep us going for a while, but working out where we are is a bit messy; too many things will have gone wrong even before we worked out what 'Friday the 2↓' is. So let us try some other way.

(It was from just such attempts that Maths developed; it did not come pre-installed with Adam.)

It will be easier to know where we are if, after completing the first 0-7 run we return to 0 to begin another 0-7 run. The problem is that these pictures do not show any signs of wear from repeated use... So, one cannot tell, just by looking at them, whether we had already been through them. Therefore, in order to distinguish the second run from the first we need, as before, to choose some indicator to show that they belong to the second run.

So far so good. But, gifted as we are with enormous foresight, something tells us that before long we will run out of pictures again. And again and again and again. So what we need is a method of indication that also tells us how many times the pictures have been repeatedly used in a lengthy count (Otherwise it will be like my son accounting for the disappearance of 180 biscuits by saying ninety times "I had only two").

So, why not simply use the same existing pictures also as the indicators that tell us how many times we have been through the series.

We must decide where to put the indicator: above? below? behind? No. In front. Why? One does not ask such questions in modern art. Because it does not matter. What matters is that everyone does the same.

How, exactly, should this work?

An intuitive way is to put a '1' in front of each picture in the 1st run: 10, 11, 12, 13, 14, 15, 16, 17 (eight numbers!), then put a '2' in front of each picture in the 2nd run: 20, 21, 22, 23, 24, 25, 26, 27, a '3' in front of each in the 3rd run: 30, 31, 32, 33, 34, 35... etc.. The question is then how many we have actually counted in reaching this last figure of 35? The (not quite accurate) story would go something like: eight in the completed first run refixed by '1'; another eight in the completed second run prefixed by '2'; finally, five in the current, incomplete run prefixed by '3': altogether, 2x eight + 5*.

This is not nice: **35** standing for only **2x** eight +5... - looks like an overstatement...

If, however, we made the prefix/indicator stand, not for the number of the current run, but for the number of runs already completed before beginning the current one (in which we reached 5 in the example above), '35' will now stand for 3 completed runs of eight steps plus another 5 steps.

In total: **3 x** eight + 5, a much nicer meaning for a number prefixed by '**3**'.

In the 1st run, which is not preceded by any runs, the prefix should be 0. Or simply, have no prefix.

When a number contains several 'pictures', each is, customarily, referred to as a digit.

As the digit on the right specifies how far we have gone in the current, incomplete run, it is advanced to the next 'picture' with each step in the count, so we call it the counting digit. The digit on the left we could call the run counter ('run' through the 'picture menu'), and let us remember - it does not tell us which run we are currently in, but how many runs have been completed before starting the current one.

If this feels unnatural, note that you have done this all your life in another context: when you are asked how old you are, and you answer 'five' (your age being 5 years and 4 months, say), you are actually somewhere in your sixth year, (4 months into it); five is the number of complete years you have lived before you are in the position to say 'I'm five' (to all those adults who can usually not think of any other topics of conversation).

We started to develop a system where the digit on the left counts the re-runs of the right (often a frustrating situation for the left...). We continue now with something that looks the same, but we change the story somewhat. (It will make things more convenient later, and also overcome the snags pointed out in the footnote above).

To save on typing, we will use now a set of four pictures only: 0, 1, 2, 3. Also to save typing, we will refer to the number of available pictures (4 here) as the *menu length.*

We already noted that however many 'pictures' we have, the first 'picture' is not *one* but *zero*, the second 'picture' is then the *one*, and as a result, the last picture always stands for one less than the menu length (the number of available pictures): When we have ten 'pictures' the last one - the tenth - is not *ten*, but *nine* (check it if you need: 0 1 2 3 4 5 6 7 8 9 - there are ten pictures), and in our four picture set, the last one is 3.

---

* If you noticed that the *35* was in fact the *sixth* item (of the 3rd run), not the *fifth* item and that in the 1st run we included the first item which stands for *zero* which should not be counted, **and,** if this also *bothered* you, you should be complimented. In the event, the two errors cancel each other out, but the detailed story is cumbersome, so we will find a better way to *present* the same method of writing the numbers.

How then do we count on? And in particular: what is the first step (after the 0, 1, 2, 3) ? As the last picture fell one short of representing the menu length, the *next* thing we write will have to stand for this menu length. So, why not simply write "menu length" after the 3? Because it is too long, and in any case, the whole thing is about not being allowed to write anything other than 0, 1, 2, 3. But as we already know, just reusing any of these will cause some deja vu. So, we use a new position. To the left. (We already know why.)

From our 'pictures', we choose the 1 and put it on the left, to say: **1 full count of the 'pictures'**, i.e. 1 menu length. Next to it, in the (original) position on the right, we restart the run through the 'pictures' - with the first picture, the zero. It says: "and nothing else" (other than the).

What we get then is:

0

1

2

3 - one short of the menu length

1	0

one menu length (=four) and nothing else

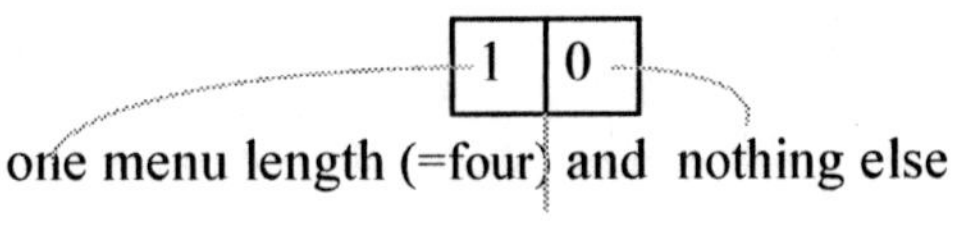

If we now use the full set of the painstakingly invented ten 'pictures', the first run takes us to 9.

What we then get is:

0

1

2

:

:

8

9 - one short of the menu length

1	0

one menu length (=ten) and nothing else

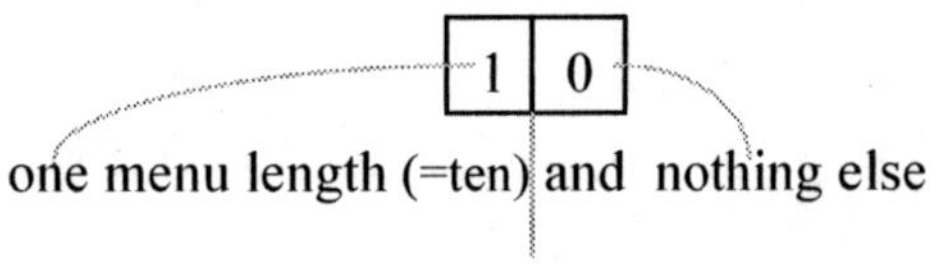

That is why the two-digit 'display' 10 stands for ::::: when the normal ten-picture menu (0-9) is used. (About time we knew this. In fact, in an ideal world this should be explained to children upon their very first encounter with the 10).

Remember: this '10' stands for the number of available 'pictures' no matter how many there are (because this is always what comes after the last 'picture', and the last 'picture' always stands for one less than the number of pictures, because the first one is always zero).

After the 10 comes 11, 12, etc., the 1 on the left reporting the (one) full count of the menu length, and to this are added the values of the pictures that are re-run on the right. For instance (with the four-long menu of 0 - 3): 12 means one menu length, namely: four + 2 (i.e. :::).

When 13 is reached, the counter on the right is, again, unable to go on and present the (now 2nd) full count of the number of pictures. To do so, we again need to resort to the digit on the left, which, now, will have say: two menu lengths. How? Quite simply, by putting a 2 there; and next to it, on the right, a 0 saying "and nothing else"; this 0 also, conveniently, restarting yet another run through our 'pictures.

What we have then, is: 2 0 meaning two menu lengths (eight) and nothing else

and 2 1 means two menu lengths (eight) + 1

So, we discover that, (and understand why), there are various ways of writing this quantity ::::. ; When we use our normal (ten 'picture') set, it is long enough to have a single 'picture' for this, the 9; But if we have only the 0 1 2 3 set, the same ::::. is written as 21 .

From there on, the count continues as:

22 -two menu lengths (eight) +2

23 " " " +3 (=11, using the normal ten-picture set)

Here, again, the last picture was reached on the right, again one short of reporting the (now 3rd) complete count of the menu length. Again, this job is left to the left - increasing the 2 there to 3:

30 i.e. three menu lengths (twelve) and nothing else (0, restarting on the right)

This is followed by:

32

33 three menu lengths (twelve) +3

What happens now?

The right, again, has exhausted itself before the next count of the menu length can be completed, but this time 'next' means fourth time, and this task even beats the left digit. It, too, cannot use a (single) 'picture' to portray this fourth time.

Let us describe exactly what has happened. The digit on the left counts the menu lengths, but this menu length counter, itself, stops short of reaching the value equal to *the menu length* (four).

So, how can we portray a counter on the left that, itself, now needs to equal the menu length?

(As typical for overspending what we have) we move even deeper into the left. To the left of our *menu-length* counter we 'open' yet another position, (the 3rd), and report *there* that one full menu-length-worth (of menu lengths) has been counted; we do so by putting a 1 in this 3rd position. So, while 3☐ said "3 menu lengths", the 1 in the new 3rd position

i.e. 1☐☐ says:

1 "menu-length worth" of menu lengths.

Now, we never forget that 'of' means 'times', so, the above means (1) menu lengths x menu lengths, or: (1) $(\text{menu length})^2$.

When menu length is four, as in the present case, the 1 stands for four^2, but with the normal 'ten-picture' system, this would be ten^2 i.e. one hundred in common English… That is why.

What then do we put in the 2nd position? There are two things we want to do, and, conveniently, they happen together: we want to restart the run of the 'pictures' there too, just as we do on the right after having reached the last 'picture'. So, 0 goes into the 2nd position. This also satisfies our other need, namely, that as far as counting menu lengths is concerned, the 1 in the 3rd position gives the full current account (of four), and so, the second position should say 'and nothing else'.

(As the [1][ ][ ] came about after 33, i.e. after a 'last picture' also on the right, we need to restart from 0 in that position, too).

So, the 33 is followed by

100.

This is how things go in the 'four picture' system (0 1 2 3). In the ten picture system (0-9), it is the 99 that is followed by 100. Knew that? Probably. But now you also know the story behind it, and from it, you can now do all this with 'picture' menus of any length.

How do we continue from here?

We always advance the picture in the right hand position whenever we can. (The other positions are advanced only when forced to do so by a 'run out' in the position to their right). So:

101

102 - 1 $(\text{menu length})^2$ + 0 (' no further') menu length counts + 2

103

As before, the position on the right is unable to portray that, in the next step, the count of another menu length will be completed, so the 2nd position must show this, by advancing one 'picture' (which, now it is again able to do); and of course, back to 0 on the right.

110, followed by

111 - 1$(\text{menu-length})^2$ + 1(further) menu-length count + 1 i.e.

$1 \times \text{four}^2 \quad + \quad 1 \times \text{four} \quad + \quad 1$

112

113

120 etc.

:

132 - 1$(\text{menu length})^2$ + 3(further) menu length counts + 2 i.e.

$1 \times \text{four}^2 \quad + \quad 3 \times \text{four} \quad + \quad 2$

133, and now?

Here, in the 1st and 2nd positions (going from right to left, remember), we again reach the situation we had in 33 above. Again, the menu length counter in the 2nd position is itself unable to reach the size of a menu length. So, again, we need the help of the 3rd position to register a second 'menu-length worth' of menu-length counts. We already had a 1 there, reporting 1 'menu-length worth' of menu lengths, so a 2 in this position will mean 2 'menu-length worth's' of menu-lengths, i.e.

[2][ ][ ] means $2 \times (\text{menu length})^2$.

As before, when the 1 had entered this 3rd position, nothing else needs to be reported at this point, and so, conveniently, in both the 2nd and the 1st positions on the right, we restart the runs through the 'picture' menu - with 0:
200.

How much then, would 231 be?

Simple: 2 (menu length)^2 + 3 (further) menu-length counts + 1. This, with menu length of four, is 2 x four^2 + 3 x four + 1;

It is a short and rewarding task to list now all the numbers from 1 to 231 - using the 'four picture' set. Then, count them (in the 'normal' way), to check that you get (45).

Remember, you list the numbers as you did until yesterday, but when you reach 3, you do what you used to do upon reaching a 9.

Now, the method that was described here works, of course, with any length of 'picture menu'. Also, happily, with the normal 'ten picture' system. Here then, 365 means, as before:

3 (menu length) ^{2} + 6 (further) menu lengths +5, and with the menu length being ten,

this amounts to 3 x ten^2 + 6 x ten + 5 i.e.

3 hundred (and) 6ty (and) 5. Good to know.

We still need to go a little further with the four-picture system.

First, let us summarize the pattern which evolved:

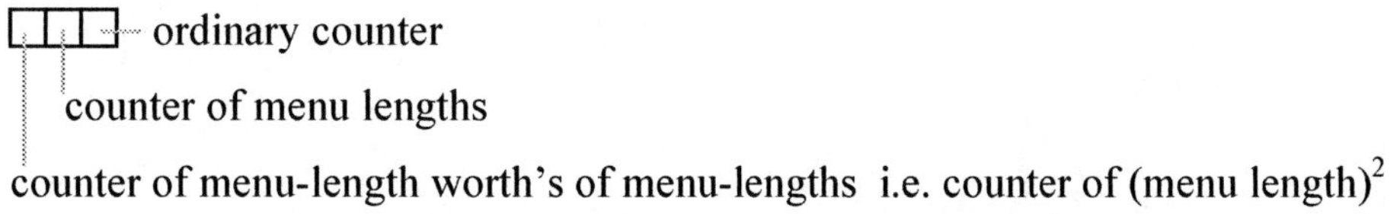

□□□ ordinary counter

counter of menu lengths

counter of menu-length worth's of menu-lengths i.e. counter of (menu length)^2

What happens after 333? (Note: it is the same as happens, in the 0-9 menu, after 999, but hopefully, faster).

Before, whenever the 1st and 2nd positions on the right reached 33, they returned to 00, and the 3rd position was incremented by one, i.e. 0 33 133 233, but: 333

↓ ↓ ↓ ↓

100 200 300 ?00

Now, this digit in the 3rd position (the counter of "menu-length worth's of menu- lengths") itself needs to be incremented to the value of a menu length, and like the others, it cannot do so. Fortunately, the left will go to any length for larger and larger numbers (provided someone else pays). So, the 3rd position will now, itself, do nothing (i.e. return to 0), and pass on the job to the 4th position - i.e. the job of portraying the required (take a deep breath:)

.... one menu-length worth of menu-length worth's of menu-lengths:

by putting a 1 in this 4th position i.e. *1* 0 0 0

Replacing *of* by *times* in

we get ...one menu-length x menu-length x menu lengths,

so, the meaning of the 1 in the 4th position can be condensed into the much neater

...1 (menu length)^3,

(with a menu length of four, this is equal to what is normally written 64).

What is this 1000 in the normal 0-9 system? With *menu-length = ten* the above means ten x ten x ten, and they have a word for it: thousand. Another that is why.

So, after the 333 (in the four picture system) came 1000, and from there on the 1 in the 4th position remains while the positions on the right go on doing what they did before, namely: 1001, 1002 etc.

Eventually, 1333 is reached, and it is the same story as after the 333: the 4th position must *again* take on the reporting of a "menu-length worth of menu-length worth's of menu-lengths" - now for the second time, (and the other positions 'resting their case' with 0's). So, after the 1333 comes

2000 - the 2 in the 4th position standing for 2 x (menu length)^3.

Incorporating this with what we already know about the meaning of digits in the first three positions:

2312 (menu length 4) means: $2\times 4^3 + 3\times 4^2 + 1\times 4 + 2$ (182 in normal 0-9).

With a menu of five 'pictures'(0-4),

...the same 2312 would mean: $2\times 5^3 + 3\times 5^2 + 1\times 5 + 2$ (332 in normal 0-9).

Instead of writing this for a specific menu length such as the 4 or 5 above, we can write it in a 'generalised' form i.e. for any menu length ('ML' for short, or even shorter, just 'L'):

$$2\times L^3 + 3\times L^2 + 1\times L + 2$$

$$2\times 1$$

Now, we can rewrite the above in a way that introduces a 'uniform pattern' in the *num.mults*:

- remembering (from chapter 3) that L^1 means just L:
- " " " that L^0 is 1:

i.e. $2\times L^3 + 3\times L^2 + 1\times L^1 + 2\times L^0$

We can further generalise this by replacing the specific digits (like the 2,3,1,2 in the above) by letters which represent any digit. We could use a,b,c,d, but it would look more uniform if we wrote d_1, d_2, d_3, etc. Note: that the d_1 is the one on the right, (- we started on the right, the 2nd digit we put to the left, the 3rd further to the left etc). So, the above then looks like:

$$d_4\times L^3 + d_3\times L^2 + d_2\times L^1 + d_1\times L^0$$

underneath each term we put its number: (4) (3) (2) (1)

Note: the subscript of the d is the same as the term's number, and the num.mults of the L is one less than the term's number. If we now also generalise the term's number into n, say, then we can refer to any of these terms as the nth term, and write it as:

$$d_n\times L^{n-1}$$

Lastly, we try to save having to write these terms four times over, with the '+'s in between:

Instead, we could write:

" The sum of terms like $d_n\times L^{n-1}$, with n going from 1 to 4 ".

Now, a *great* shorthand: for we draw a '$\sum$' and to denote the we write $\sum_{n=1}^{4}$.

Looks awful but is short.

So, the above can be shortened into: $\sum_{n=1}^{4} d_n \times L^{n-1}$. Scary, but only to others, right?

Whenever you encounter such monstrosities, don't panic: these are mere short-hand for very simple concepts. All you need do is to slowly remind yourself (or ask, look up, etc.) what these 'drawings' were chosen to abbreviate. Say you found a pack marked A.T.E; you wouldn't just run, you would simply ask what it stands for. (You might be told: "About To Explode", and *then* you run. But only from the bomb, not from the shorthand.)

($\sum$ is the Greek letter for S, as in Sum).

Let us practice: 587 with menu length 8 i.e. a system of 8 pictures, (a common diet for computer programmers). What is this? A mistake, is what it is! An 8-digit system has no 8: it runs only to 7 (because it starts with 0). So, let us take 527 instead.

In our $\sum_{n=1}^{3} d_n \times \mathbf{L}^{n-1}$ n goes from 1 (on the right) to 3 (on the left), so $d_1 = 7$, $d_2 = 2$, $d_3 = 5$

L is 8; when n is 1 then L^{n-1} is 8^0 (=1)

" " 2 " " 8^1 (=8)

" " 3 " " 8^2 (=64)

With all these:

$$\sum_{n=1}^{3} d_n \times \mathbf{L}^{n-1} = 7 \times 1 + 2 \times 8 + 5 \times 64.$$ This, using the normal 10 picture menu, is 343.

4.2 What you always knew, but now also understand

(7) (4) (1)

For example in 3426795 the *6*, nesting in 4th place, contributes 6x 10^3 (6 thousand); the *3* in the 7th place contributes 3x 10^6 (3 million) (*million* is 1000 x1000 = 10^3 x 10^3= 10^{3+3} = 10^6, but you knew that anyway). This is why we write numbers in the way we do to represent the quantities we wish to express.

4.3 More ways than one to skin a cat

You have seen that the quantity ::: ::. (normally written as *11*), can also be written as *23*, and a quantity that we normally write as *45* can be written also as *231*. This we understand now. However, a problem does arise (as usual) when we begin to *talk*.

In the above situation, what should we *say*? "forty five" or "two hundred thirty one"? Why not simply say what we mean, namely, say where each of these are coming from?

We 'drew' the *231* when we wanted to express a certain quantity while having a menu of four 'pictures' (i.e. *menu length 4*). The common (yet, acceptable) phrase for this is 'base four'. (Yes, '*base*' = '*menu length*'). So, one could say here: '*two hundred thirty one, base four*', and write it as $231_{(4)}$.

We 'drew' the same quantity as *45* by using our normal sumptuous set of *ten* pictures (called the *decimal* system, from the Latin *decem* meaning ten). So, we could call it '*forty five, base ten*', or *forty five, decimal*, and write this as $45_{(10)}$.

When we hear or see a number without any specification (of the base) we assume it is the one that is normally used, namely, *base ten.*

Another question that needs clarification: When we **say** *ten,* do we mean the *quantity* ::::: or the *'graphic display' 10* ? If the latter is *decimal* we need not worry, but if, when **saying** *ten* we refer to the *'graphic display' 10* and this *'10'* is not decimal (i.e. the menu length is not 10), we should always say "ten, base so and so", and this clarifies that we mean a quantity that is *not* ::::: (it is, of course, equal to the non-decimal *menu length* that is being used).

To ascertain that you can now make out numbers, given in any base try the following: How much is *twenty four, base seven* i.e. $24_{(7)}$? The *'menu length counter' 2* on the left says *2*x *7*, and the digit on the right reports a further count of 4. Happy graduation.

We leave it to the reader to verify that $41_{(5)}$ is the number of days a hen needs to make eggs squeak.

4.4 This is now childishly easy, but without it don't even ask how computers work

What, one may ask, is all this good for, that is, when and why need we ever venture away from the safety of the *decimal* system? (To ask such questions is a sure sign of recovery from the slave mentality of school maths.)

Any activity in a calculating machine involves constant switching from one number to another, and the faster it can do this the faster you finish your homework.

In the days when toys were made of wood, calculators had wheels with digits on their rims. A seven digit number, say, was presented on a set of seven such adjacent wheels, using one digit from each wheel, as in car mileage recorders. To change the number the wheels had to be set in motion and stopped again, and this could not be done faster than a few times a second. Calculating machines working at that rate could barely forecast last year's weather. Nowadays if your computer switched numbers less than billions of times a second you would deny it was yours.

How does a calculator do this?

The reason calculators are called 'electronic' is that one can switch electronic components from one distinct state to another billions of times a second, because electrons can be activated more easily than wheels. An electronic component can be in any of many different states, for instance by carrying different levels of current. But to identify these states they have to be accurately measured, and always accurately maintained, which is both costly and slow. It is much easier to have only two, easily distinguishable, states, for instance 'on' and 'off'. Then all that matters is whether it is on or off; there is no need to measure '*how* on' it is. This lavish assortment of just two states is what a computer is equipped to count with, and that is all it needs.

Really?

Indeed, we will now set out to count in a way we never imagined we could, this time using only a '2-picture' system. This is simply what the show-offs call 'Binary' system when they want to intimidate you, but now you have caught up with them.

First we must decide what 'pictures' to use. We could invent our own. However, to avoid the need for expensive graphic artists, we will just use the first two of the already existing 'pictures' '0, 1' - the first 'picture', as we know, must be 0, the second, and already the last, is 1.

(Inside the computer it does not matter whether 0 stands for 'current off' and 1 stands for 'current on' or the other way round, as long as one remembers which way round it is when the computer is connected to the outside world.)

In the following, we will write the '2-picture' numbers on the left; to their right we give their decimal equivalent. Further to the right: illuminating comments.

('*o*' stands for a front-end 0 which *can* be written but usually isn't; '0', like a tail, has no use at the front, and just looks odd.)

We proceed by going through the series of digits in the usual way, except that whenever we reach 1 we act as if we had seen a 9, by returning to 0.

ooooo0	(0)	the obligatory prelude to any counting system
ooooo1	(1)	What next? *1* is the end of the series: act as you would after seeing a 9 there, i.e. return to 0, and in the next position to the left, advance the *0* to *1*:
oooo10	(2)	Note: the decimal value of this *10* equals the *menu length* (number of 'pictures')! Next, advance the rightmost digit (now you can):
oooo11	(3)	Regard this as having the same 'future prospects' as a *99*:
ooo100	(4)	Note: the decimal value of this *100* equals the (*menu length*)^2. Carry on in the two rightmost columns as in the previous four steps, but now with a *1* in the 3rd col. :
ooo101	(5)	
ooo110	(6)	
ooo111	(7)	Next, act as if you saw *999*:
oo1000	(8)	Note: this equals (*menu length*)^3. Now repeat lines 0-7, putting a *1* in the 4th col. until you get to *1111*, after which...
o10000	(16)	Note that in the 10-picture system *10000* is 10 times bigger than *1000*; in the 2-picture system *10000* (=16) is 2 times bigger than *1000* (=8) Since this is such fun, why not carry on by yourself and do a few more...

This is all becoming now pleasantly clear and therefore unpleasantly boring. Clearly, this can go on much longer; in fact, like the shorter Mahler symphonies, forever. With a series of just two 'pictures' one can get just as far as with the normal series of ten. The only difference is that the displays are longer: we have seen that the decimal 16 appears as 10000 when a menu of only two pictures is used. (The fewer the pictures, the larger the exhibition…)

How much, for example (in decimal equivalent), is *1110011*?

We now know the method for working this out, since we know how to work out the value of numbers arrived at by using *any* menu length; that is, you will recall (with a smile):

$$\sum_{n=1}^{T} d_n * L^{n-1}$$ (T is the number of digits, 7 in the above example)

So with *L* (the *menu length*) = 2 as in the present minimalist system, this is

$$\sum_{n=1}^{T} d_n * 2^{n-1}$$

We note with delight that here d_n can only be 0 or 1, both so easy to multiply by.

In working out the value of 2-picture (binary) numbers it helps if you prepare a list of *Selfmult* 2 with higher *Num.mults*, that is beyond the basic $2^0=1$, $2^1=2$, $2^2=4$, $2^3=8$.

The list goes on as $2^4=16$, $2^5=32$, etc. .... $2^{10}=1024$

In working out *1110011*:

we start from the right and proceed to the left (like getting younger…):

1 in position 1 contributes *1* x 2^0 which is just 1

(remember - the *num.mults* is one less than the position no.)

1 in position 2 contributes 1 x 2^1= 2

0 multiplying anything in positions 3 and 4 provides a welcome respite before multiplying the

1 in position 5 by 2^4, which contributes 16, and the

1 in position 6 which contributes *1* x 2^5= 32, and

1 in position 7 which contributes *1* x 2^6= 64.

Totting them up we get 1 + 2 + 0 + 0 + 16 + 32 + 64= 115. (Note the order of the contributions: always from the right to left)

(If you don't try some of these by yourself you are a masochist…)

Have you heard about *k*? The Romans asked us to use *kilo* for 1000 and we agreed because it is shorter than *thousent* and easier to get the spelling right. Computer folk don't even bother about *kilo*, they just use *k*. But what they mean by it is 2^{10} which, as you may have noticed above, is only *approximately* 1000. To be precise it is 1024. So, when '*k*' is used in the context of computers it is in fact 1024, not 1000, and people who know this are rated as being very clever. You too now.

And therefore you are ready for something really nice:

4.5 And now for the Winner

We said that this would be nice, but wait until you find out *how* nice: it is the basis of a method of betting that *guarantees* that you win every time. (I know, you have just remembered that you really always *did* like maths…)

This is the way it works: You start with a bet of £1. A die is thrown. If a five comes up you win; if any other number is thrown you lose. So far so bad: the odds are 5 to 1 against you. You play on, doubling your bet at each throw. In return for accepting such unfavourable odds you are at least allowed to decide when you have had enough. So you begin the game and go on losing, but you go on, because you know that eventually 5 *will* come up, no matter how long it takes, and then you can pack it in and leave with a net win! How come?

Let us suppose you suffer seven losses before a five turns up. You will have lost £1, then £2, £4, 8, 16, 32 and £64. When you win the next throw it is £128. But how can you be sure in advance that your win is bound to be greater than your accumulated loss?

Because you know so much maths.

Your losses - 1, 2, 4…., 64, are all *self-multed* 2s with *num.mults* 0, 1, 2 etc., i.e. 2^n with n going from 0 to 6 (this is *7* steps, and remember: $2^0=1$)

The *sum* of those losses can then be shorthand-written as $\sum_{n=0}^{6} 2^n$.

For a reason that will become beautifully apparent, we want to rewrite the above as:

$$\sum_{n=1}^{7} 2^{n-1}$$ (***check carefully: it means exactly the same***)

Now, does this not look *just a little* like the meaning of the 2-picture (binary) number *1111111*?

It could not look much *more* like it, because that is precisely what it *is* (allowing for the absence of the *d* x*'s* which can be omitted here because in this case they all are *1*x).

Now to your *win*: it is twice the last loss of 2^6, i.e. $2 \times 2^6 = 2^7$.

As we discovered that the accumulated losses can be expressed as a binary number, it makes sense to try and express the win, too, as a binary number.

Indeed, the 2^7 can be written as $\sum_{n=1}^{8} d_n * 2^{n-1}$, provided d_8 is 1, and all the other (7) *d*'s are 0.

Why? As only one term in this sum survives, the Σ plays no role. In the only surviving term, the one where *d* is 1(x), n is 8, i.e. n-1 is 7, so the 2^{n-1} is 2^7, our win.

What binary number is this? *1* in the 8th place and the rest *0* i.e. *10000000.*

How does one get to this number? When we climb to *10000000*, the last thing we see before reaching this nice round figure is the spiky *1111111*, the sum of our losses.

So, our win is 1 more than the accumulated losses. This will be the case no matter how long it takes for the win to arise.

Worth trying out? ***You bet.*** But hurry, before everyone has read this book, and no one wants to play with you…

4.6 A look behind the point

Let us now move on and look at the little things on the far right.

In the decimal system, each time 10 steps are made in the first (rightmost) position, only one step is made in the position to its left (the 2nd position), and for each ten steps in this second position only one step is made by *its* neighbour to the left, the third position, and so on.

This means that one step in any position represents something that is ten times smaller than one step in the next position to its left. For example, one step in the 3rd position represents a *hundred* which is ten times smaller than one step in the 4th position, a *thousand.*

The smallest, the *unit* in the first position on the right has no neighbour to its right, and gets cold. So let us give it a neighbour to its right as well. It will no doubt be very grateful for this, however it is not prepared to relinquish its 'first position' or 'unit counting' status! So we insert a mark between it and its new neighbour to signify that it was, is and always will be the first and only unit counter. We could put an *f* for *'first'*, but the budget allowed only for a dot, a *point*, as it is called, e.g. 543.

Now, this new neighbour is not only there to keep Unit Counter warm; it is there to serve us.

Of course it must obey the rules of the other members of the family it has joined, so it too will have to make ten whole steps to move Unit Counter on its left one step, let us say from 5 to 6:

5.0 5.1 5.2 5.3 5.4 5.5 5.6 5.7 5.8 5.9 6.0. Each step results from an increment of 1 in the digit to the right of the point, so we write this increment '.1' and call it *'point one'*. The word for this quantity, of which ten are needed to make a whole unit, or, conversely, the quantity which we get when we divide a unit into ten is *a tenth,* which is written $\frac{1}{10}$, or... yes, yes...10^{-1}.

That is *why* .1 is the same as $\frac{1}{10}$!

If we go only *half* way through the series of the ten .1 steps we get to 0.5 - this gets us only *half* the way to a full unit (1). That is why .5 is half, or ½. Good to know why, at last. Isn't it?

If this tenth is also to have a neighbour on its right, it will have to be one that also conforms to the same rule and go through the series of ten steps for the digit on its left to grow by one step (of one tenth), for instance: from .3 to .4. This progression will then appear as .30 .31 .32 .33 .34 .35 .36 .37 .38 .39 .40. Each step results from an increment of 1 in the second place behind the point. We write this increment '.01' and call it 'point o one'.

But what is this thing, ten of which make up a tenth? Or conversely, what is this thing which we get when one tenth is divided into ten?

We know four answers to this: three nice ones, *a hundredth, 1/100,* and 10^{-2}, and one that only exists, unnecessarily, to confuse us: *one per cent.*

So 684.72 means:

6x hundred (10^2) + 8x ten (10^1) + 4x unit (10^0) + 7x tenth ($1/10 = 10^{-1}$) +2x hundredth ($1/100 = 10^{-2}$); or:

6x 10^2 + 8x 10^1 + 4x 10^0 + 7x 10^{-1} + 2x 10^{-2}. (Note the progression of the *num.mults*)

What would be the meaning of 1 step behind the point in a 3-picture system (0, 1, 2)?

In this case each position advances only three steps to make its neighbour on the left move. The same therefore holds for the position to the right of the point: 3 steps of 0.1 will be needed to advance the units position. For instance the progression from 1 to 2 will appear as 1.0 1.1 1.2 and we are already there: 2.0.

Now, what is this thing of which we need 3 to make up one unit?

A *third* or 1/3. So that is what .1 is in the 3-picture system. And .2 is *twice* this, i.e. 2/3.

So, in the 'normal', decimal system, .1 is 1/10

in the 3-picture system .1 is 1/3

so for a system of any number of pictures .1 is 1/number of pictures (or 1/*menu-length*)

What then does .1 mean in the 2-picture or binary system?

Half an answer will never do, but *half*, here, will do very well as an answer.

4.7 Graduation

Should there still be any doubt about the importance of understanding this chapter, an interesting observation may be made.

You now know how to count on and write the next seven steps after '5545' in a 6-picture system, and how to work out whether you would accept that much for your car. Before, when you had no idea how to do this, you were still able to write things like '1995' and work out how many years would pass before the big party. You knew how to use numbers correctly, but not *why* they are used in the way they are.

This was similar to the, hopefully by now defunct, *show and tell* method of teaching children to read and write: display "*parrot*" on the board and recite what this word should sound like. Trying to get them to read and write without disclosing the method - making them do what parrots do ...

Children eventually figure out the phonetic rules, and they are then also able to use these rules in 'different systems', that is, different languages. But with numbers the situation is worse: even people who, all day long, use numbers, dream numbers, pray numbers, do so in a parrot fashion, they do not really *understand* the method. This is revealed, for instance, by their total inability to use non-decimal systems.

You are now, numerately, de-parrotised.

Chapter 5

An Antidote to Paranonsenseology

5.1 Familiar territory in which most get lost

When we hear, speak and think about the *probability / chance / likelihood* of something, what do we mean, what should we mean, and how can we evaluate its correct magnitude?

It is a particularly important to study this topic in maths. Those who do not know what *calculus* is about do not use it, but everybody messes about with probabilities. Yet this is one area, like nuclear or genetic engineering, where doing it wrongly is worse than not doing it at all.

By 'wrongly' we do not mean minor inaccuracies or deficiencies, which are usually inevitable anyway. We mean misunderstanding the concepts involved, and performing apparently correct operations without understanding where they come from. This is different from not understanding the reasons behind arithmetic rules, where at least no harm is done as long as they are used correctly. Misunderstanding concepts of probability breeds and nurtures a whole family of beliefs many of which are prefixed by some *para-*. As misguided hobbies they are inconsequential, but when they lead to wrong decisions it gets serious enough to warrant some learning.

There is a vast amount of material about mathematical methods for dealing with probability and statistics. We will not go into too much of this. Rather, our purpose is to dispel some gross misunderstandings in this area. In the process, we will however learn one simple and interesting mathematical tool, without which no evaluations of probabilities can be attempted.

The purpose of everything here is usually to do with predicting the future.

There are four types of situation regarding systems the future performance of which we try to predict.

a). Where we *understand* how the system behaves

'Understanding' means knowing why a system acts the way it does. Given that, we can establish with certainty what it will do in *any* given situation. For example, with the right knowledge of physics we can always predict where a thrown stone will land, even if we launch it with a specific velocity we have never tried before. Therefore there is no uncertainty here, and thus no job prospects for statistics/probability.

b). Where we *appear* to *know* how the system behaves

By 'knowing', rather than 'understanding', we mean being in possession of a rule - a description of an accurate pattern of the system's *past* behaviour. This implies that we are probably safe in predicting events in circumstances very similar to those in which the system's operation was observed in the past. But as we do not know *why* it operates as it does, we cannot apply the rules with much confidence in situations which are different or potentially different.

c). Where we have partial knowledge and / or understanding

Where events appear to follow the rules, but only approximately.

d) Any of the above situations where, even if in principle, we could make predictions (more accurately in (a), less so in b and c) but so many elements and steps in the process are involved that the predictive calculations are not feasible.

5.2 Generally required procedures

In situations b, c and d there exist theories of probability and statistical methods that make predictions more than just guesswork. Generally the procedure consists of the following three steps:

1. Producing some assumed rules. These could derive from partly theoretical considerations, in some cases from probability theory itself, or from intuition, or by observations of actual situations, or from opinions or reports typically gathered from surveys.

2. Applying certain *statistical methods* to establish numerical values that best make the rule fit the reality.

3. Applying 'statistical tests' to verify the validity of these rules.

What is meant by this is that if a proposed rule is found to fit all past experiences, and it correctly predicts any previously untried situation, the theory is correct. This is not necessarily the same as 'understanding', because we do not know *why* it works, but it is effectively so for the purpose of most immediate predictions. In this case we do not need the help of statistics.

If the rule does not tally at all with past *and* current experience it is scrapped.

If the rule partly fits, i.e. the actual events are scattered around the predicted ones, 'statistical tests' are used to try to establish whether deviations from the rule are pure chance, or whether they indicate a systematic discrepancy. The tests also tell the degree of reliability one can expect of predictions from the rule, i.e. what proportion of actual events can be expected to fall within a given range from the prediction.

Statistical tests are developed from the 'theory of probability'. This theory deals with establishing expected relative frequencies of compound events where the basic constituent events are considered to behave randomly.

All these generate predictions that are by no means certain, but are likely to be closer to the actual future outcome than a purely guessed forecast would be.

Let us now consider one very simple case in which we try to make a prediction in the face of uncertainty.

We throw a die. Since any one of its six faces is equally liable to turn up, we do not know which one actually will. But we still want to predict the sum of ten throws.

With certainty we know only that the sum must be at least 10 x 1 and at most 10 x 6, but in between these we do not know what the sum will actually be. Can we do more than just guess?

We would not expect the die to always turn up 1, nor even always turn up a small number. Similarly, we do not expect only large ones. We would expect small numbers at some throws, large numbers at others. Therefore the sum of a few small numbers and a few large ones would resemble the sum of all medium numbers.

What we mean here by 'medium number' is the probabilistic notion of *average*. In this case, with 1 as 'small' and 6 as 'large', the medium, i.e. average, is half way between 3 and 4, i.e. 3.5, which is no closer to 1 than to 6. (It does not matter that a die cannot turn up a 3.5; what matters is that if it turned up a 3, then a 4, the result would be the same *as if* in both cases it turned up 3.5). So we would expect the sum of 10 throws to be somewhere near 10 x 3.5 = 35.

This still does not tell us much about what will *actually* happen if we try it; try it once, that is. But try it *many* times, i.e. many sets of 10 throws, and tot up each set, and you will indeed find most of the sums to be near 35. It is worth a try.

As a tool for performing predictions we use *models*. This is often a reduced physical replica of the real system (hence its name), or it may be a recipe-like list of instructions containing the rules for a multitude of step-by-step operations of many participating elements. The number of elements and/or steps can be very great, in which case the list had better be a computer programme. Running this programme is called 'computer simulation' of the system.

In its most common, and believe it, most convenient form, this 'recipe' comes in the form of a mathematical expression which can contain both the known (or partially known) rules and the 'statistical supplements'.

By *probability* we express the degree of likelihood that we attach to something being in a particular state rather than not being so, in the belief that we have reasons for assuming that likelihood. We express the degree of likelihood as a fraction of 1. This means the 1 is split into two parts; the size of the first part, called 'the probability', reflects the assumed likelihood of something being in that state; the other part reflects the assumed likelihood of it not being so.

More will be said about what, exactly, is meant by the number (the fraction) that expresses the probability.

This fraction of 1 is commonly expressed with 100 as the divider; for example a fraction of ¾ is expressed as 75/100, i.e. 75%.

A probability of 100% is thus the same as a probability of 'a full 1', leaving no room in the 1 for the other part that represents the possibility of 'not being so'. Thus a 'probability of 100%' means certainty.

Probabilities, though, generally deal with situations of *un*certainty. This means that when we are considering the probability of an occurrence of an event at a particular instance, we do not know, at least not fully so, what causes the event to occur or not to occur at that instance.

There are two interesting and important points in this connection, namely, regarding the relationship between probability and reality. They involve two common misconceptions:

- the belief that unpredictability implies that events are not necessarily fully *caused*

- the belief that uncertainties and statistical trends are inescapable fate

These need to be cleared up because they have serious consequences.

We deal with the first one next, and come to the second one at the end of section 5.4

Uncertainty, or unpredictability, does not imply non-causality

Our uncertainty does not mean that there *is* no reason why the event does or does not occur. Everything has a cause; it is just that *we* do not know what it is, or are unable to calculate its outcome before it occurs.

We refer to the origin of our uncertainty as 'randomness' in the situation, but this randomness is only apparent. When a coin 'randomly' lands face up, or a raindrop 'randomly' falls on our nose, they 'know' exactly how they got there: a series of very specific physical causes. We, however do not know; and even if we knew how they work, and used all the computers in the world, we could not complete the calculations necessary to predict the fates of the coin or the raindrop before they had already obeyed their causes. There are two reasons for this. One is that too many, interrelated, factors are involved. The other, less obvious, is that in certain situations the prediction of the outcome of these factors depends on knowing their 'starting states' to a practically impossible degree of accuracy. The resulting behaviour is what is known as 'chaos'. But it is only *apparent chaos*, and should be called so.

5.3 What the important concept of (apparent) chaos is all about

Consider such a process: It begins at a particular point and ends with a certain result. You then want to repeat this run (or equivalently, predict its outcome by calculations).

With normally familiar and intuitive systems, if you restart close to the original starting point you would expect to finish close to the original end point. If, however, it ends up somewhere very different, you would conclude that this particular system is very sensitive, and so, had you simply been more careful about getting the repeated starting state closer to the original one, then you *would* have ended up close to the original result. But imagine now that, no matter how close you take the repeated starting state to the original one, you still get nowhere near the original result. When this happens, you have 'chaos' on your hands.

There are no miracles: had you repeated the run with exactly the same starting state, you would indeed get the same result repeated. The bad news is that here *exactly* means ***exactly***. *Infinitely* so! Unless you use the *actual same* starting point, same to infinite accuracy, then no matter *how close* your repeated starting point may be to the original you cannot be sure of ending at the original result, not even *anywhere near* it.

As it is, of course, impractical to measure and record the starting point with infinite accuracy there is then no way to predict confidently the outcome of such 'chaotic' processes.

Situations of this sort typically arise in systems in which 'branch points' are encountered in the unfolding procedure. If the original (or assumed) run 'hits' the branch point on one side and the repeated (or calculated) run on the other, then the rest becomes unrelated to the first or assumed progression and thus unknown.

For a substantial likelihood of the two runs encountering such branch points close enough to be split by them, the system must also be such that small deviations in the starting point are greatly amplified during the ensuing steps.

To visualise a realistic *chaotic* system, imagine many spaced layers of spaced horizontal steel rods. Two small steel balls are dropped, from *nearly* the same point, onto one of the top rods. They must bounce off either of the two sides of the rod, depending which side of the rod's ridge they fell onto (but no matter how close to the ridge). After bouncing off this rod they drop on to further layers of rods, etc.

Due to the curvature of the rods' surface, each time they bounce off a rod the gap between the two balls increases. Eventually, the gap between them will be such that they reach different sides of some rod, *and from there it is good bye for ever.* The point is that no matter *how tiny* the initial gap between them was, after some number of bounces the gap will be large enough to make it very likely for them to drop onto different sides of some rod's ridge. If, and where, this happens depends on their initial positions, *to an infinite* degree, and so, in practice, the final outcome cannot be predicted with certainty.

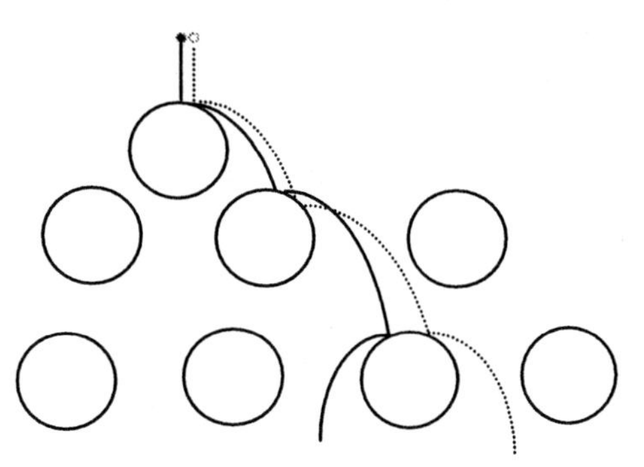

Here the 'separation' is shown to occur after three steps, but had the initial gap been much smaller the separation would still occur – many steps later.

This is how 'chaotic' systems differ from intuitive ones, in which minute differences in the starting position have only minor effects on the result.

So although to the balls it is all purely physical and causal, as far as you are concerned the outcome remains an uncertainty (until it has actually happened). All you might know is where *on the average* the balls emerge, but on any specific instance you can only wait and see.

Imagine many, very many, balls doing this, and also colliding with each other on the way. Then replace the steel balls and rods with molecules of air and water, and try to predict what these will end up doing: What *you* are doing is weather forecasting!

5.4 A salvation for reason

The most important conclusion to be drawn from the above is that although things may be eternally uncertain and unpredictable, no mysticism is required to account for this elusiveness. Situations evolve in totally causal ways. It is only we who cannot keep ahead and measure up to the task of predicting them, not only because the required quantity of the data and calculations exceeds our practical capacity, but the required accuracy defying even any *theoretical* capability.

Probabilities is what we use then to help us out in our ignorance. But they are used only as a *description*, not as an *explanation*.

(Yes, there is the issue of the source of uncertainty in Quantum Theory, but that will hopefully be dealt with in a further book...)

We now come to the second issue in the relationship between probability and reality:

The uncertainty behind probabilities is not always fate, and statistical trends are not *the* only way into the future.

5.5 Where not to use statistics, however tempting

Statistical methods are based on treating anything about which nothing is known as if it were random.

What we mean by 'randomness', essentially, is 'as far as I know *I don't* know...'

If you *do*, or *could*, understand the reasons for the way a particular event occurs, your predictions, based on this understanding, will be much better than they would using statistically based patterns or 'rules'.

This cannot be over-stressed: all too often much attention is lavished on statistical refinements while the questions of which factors should be involved, and what can be known about their behaviour, are neglected.

Worse still, the origins of models based on statistics are less transparent than ones based on understanding. There is, therefore, a danger in extrapolating such models into the future, as it is often done - without realizing that the nature of the system upon they are based can sometimes change.

5.6 Don't predict when you can design

Let us consider one common example: the prediction of changes in the economy.

This area is certainly far from fully understood, and the models are in large parts based on statistics and are heavily dependant on a small number of certain 'macro-economic indices'. This is not satisfactory because for instance, prosperous economies do not go into serious decline just "because of fluctuations in economic indices" as we are so often told; they do so, barring natural disasters, owing to somebody (or -bodies, but not only governments) doing something wrong (or just not doing much at all). Specifically, the performance of a system is governed by two types of factors:

1. The nature of the elements of which the system is composed, and how they interact.

2. Factors imposed on the system from the outside - often regarded as 'controls'.

The unfolding of economic and market situations is sometimes likened to weather changes. The two will be compared here because while in some ways there *are* telling similarities, the differences illustrate the great, but often overlooked, scope for doing more than just the things suggested by global, statistics-based considerations. In a nutshell: not much can be done to stop rain, but much *can* be done about not getting wet.

In the weather system, the first type of factors, the 'nature of the elements', is the mechanics of air and water molecules. In economic systems the first type of factor is the acts of people.

The second type, the 'controlling' factors in the weather system are solar radiation, other cosmic contributions, and perhaps ecological measures. In economics they are economic policies, disasters, and discoveries/inventions.

What the two systems have in common is that extreme changes can occur in an *apparently* chaotic way as a result of minor events which then affect the uncountable, untraceable, intractable and discontinuously interacting elements. Detailed predictions are therefore difficult or even impossible, and so, *macroscopic* tools, based on probabilistic, or statistical models are often used. (Recently, attempts have also been made at using new tools developed from 'chaos theory', which is the mathematical analysis of 'chaotic' phenomena, an example of which was shown above.)

(*Microscopic* models involve the description of the performance of the individual elements of a system; *macroscopic* models treat the whole system as a single entity, each feature of which represents the average, or sum, of the individual features of all the microscopic elements of the system.)

In using external, controlling, factors to affect systems, less can be achieved in an economy with regard to catastrophes and discoveries, and more, for what it is worth, by economic policies and planning. These typically consist of making rules - imposing restrictions and providing inducements, and of varying a small number of 'rates', for instance interest rates, exchange rates, tax etc.. These changes are chosen partly with the aid of models which represent theories about the behaviour of the system as a whole, and partly with the help of statistical methods.

As to controlling the weather... not many prospects for activity in the Ministry of Weather: a little cloud seeding here and there, some lip service to green policies (advertising that excessive combustion warms the atmosphere, thus risking that the English set alight anything that catches fire in their desperate yearning for warmer weather...), but absolutely no control over the sun and the rest of the cosmos.

For the Economy, however, (though it sometimes appears more out of control than weather...) the *potential* exists for much better control than of the weather. Here are the great differences in the nature of these two systems:

i. While we have quite adequate understanding of the physics of air, water and heat, much still needs to be understood about how people act in ways that effect the economy. This requires studies of a very different nature to those currently done to develop economic models. More detailed and imaginative factors need to be involved, and their behaviour needs to be *understood* rather than just measured and established from statistical analyses (such as apparent correlations between a small number of macroscopic variables);

ii. The even more important difference: no redesign budget will ever get air molecules to reconsider what they always do when they collide. That is to say, the behaviour of the *elements* of that system cannot be changed. But we dear people, the elements of the economy, can change, through education and communication.

Thus in the system of an economy if somebody (-bodies) were to do something wrong, they might perhaps be brought to do it right. Making use of these points must surely yield greater improvements that tweaking a few rates and just waiting for the economy to 'cycle out' of its difficulties.

5.7 Don't estimate when you can know

Let us give another, slightly more specific, example of the advantage of using a degree of understanding instead of relying on statistics only.

You are at the airport waiting for a flight that you booked, and time is of the essence. Putting no faith in assurances that 'all is fine' from the airline desk, you want to assess the chances of your flight leaving on time, with a view to possibly switching to an alternative flight or airline, if necessary.

The 'statistical approach' would be to consult records of the airlines' departures for punctuality, detailed by time of day, weather conditions, destinations and a host of other factors.

Taking the other approach, you might consider a well-known aeronautical principle: that a plane cannot take off unless it has landed first. So you go and find out whether your plane is actually there. Or, should you notice a commotion at the check-in, coming from a large crowd of overbooked passengers, and you assume that planes are not made of rubber, you might volunteer to accept the compensation for giving up your seat, book the following flight, and arrive before the original flight with its fifty overbookings had been sorted out.

Understanding an event is often confused with contrived explanations. An explanation amounts to understanding only if it has been rigorously proven.

Occasionally an event does arise that defies any explanation. Attention should then turn to examining whether the 'miracle' could arise by chance. Fairly simple considerations of probability can sometimes prove that even the most extraordinary things can in fact occur by pure chance. The problem is that, simple as these considerations of probability are, we are innately bad at appraising their result intuitively. It is this, and insufficient respect for the concept of rigorous proof of proposed explanations, which account for the readiness of many to accept a wide variety of tales, the nature of which inspired the title of this chapter.

This is the invariable rule: if an investigation of a remarkable event fails either to provide an explanation or show that it can be expected to occur by chance the investigation was *incomplete*. Any other explanation, however appealing, is deceit.

Statistical 'pattern-discovery' can indeed be used in the process of understanding, often as a source for clues. However if, as in most cases, it is not treated carefully it leads to nonsense.

This is best illustrated by using a tangible example, one that deals with very actual matters familiar and important to everybody, and at the same time shows up extreme misinterpretations of data: the frequently voiced objection to investment in roads on the ground that "more roads only generate more traffic", based on an observed pattern of increased traffic on newly widened roads. The observation is correct, but what about the conclusion drawn from it?

It is this: road improvements do not generate a desire to travel - even on widened roads people do not just ride around for fun. The volume increases because the improved road absorbs a *pre-existing* (and already paid-for) need for adequate mobility, mostly needed for productive purposes. When a town gets its first sewer and this immediately overflows, it does not mean that new sewers induce increased appetite, it simply drains away a pre-existing (kerbside) problem (graver than traffic congestion...).

Add to this an even bigger oversight: people let themselves be preached out of their cars on the 'ticket' of Ecology. Why, cars can run on fuel that leaves behind water vapour only, and that can be produced from renewable, non polluting, resources. True, this costs much more, *but UK drivers* ***are*** *already paying this cost anyway*! *In fact, much more*: using only that *minimal arithmetic aptitude* required to tot up one's car related taxes (e.g. over 80% of one's fuel bill is tax) and multiply this by UK's car population, one would discover that the result is more than *six times* the state's total expenditure on anything and everything to do with road transport. (This vast excess, of some £35 billion *annually - four Channel tunnels'* worth *annually* - is used for purposes unrelated to transport, unlike in leading Economies as the USA, Japan and Germany).

The above leads to an interesting conclusion that is not often heard: 'No more roads' is *not* necessarily the only implication of the statistical observation of 'more roads → more traffic → more pollution'. Instead, the following question should be examined: were these vast excess taxes used to fund efficient, clean and *free* mobility, (the 'other purposes' *temporarily* funded differently), might the enormous economic benefit of the resulting healthy transport system not only proceed to cover the costs of those 'other purposes', but leave *much* more behind? (*and* - make travellers life tolerable).

So, statistical trends alone (and the colour 'green') are not enough to paint the *whole* picture.

While we are on to transport-related disinformation, here is another one that feeds on the public's innumeracy: "Covering the whole countryside with tarmac" - enforced by graphic strips filling the map.

With the approx.1000km long British mainland displayed on a large, 33cm tall TV screen (from which the scale can be calculated), one *should* have little difficulty working out that an evil motorway – 24 metres wide, 3-lane each way – should appear a mere *one tenth of a human hair* wide!

(Scale: 33cm/1000km = 33cm/100,000,000cm = 1: 3 million – meaning that *all* distances on the map are 3 million times smaller that the real thing – including the motorway width:

24 metres x 1/3,000,000 = 24,000mm x 1/3,000,000 = 0.008mm = 1/10 of a *typical 0.08mm* human hair).

Putting this another way: a single one of these typically shown 3mm wide strips (≈ 400 times wider than the true 0.008mm scaled motorways width), running the length of the map, would depict not one pretended motorway but *three times* the combined area of the *entire American and European motorways networks*!

So no need for cows to diet as yet.

5.8 The meaning of stated 'probability'

Firstly we need to understand the source of the numerical values that are used to describe probabilities more accurately than statements such as 'very likely' or 'rather improbable'.

Think of some happenings, all of which conform to a well-defined description.

These **happenings** can have a **variety of outcomes.**

The concept of 'Probability' is used to describe the likelihood of a special sub-section of these happenings, namely, those happenings that have some *specific* outcome or outcome group.

Consider the probability that a matchbox that you find contains 2 matches:

The *happening* here could be: 'Finding a matchbox (with *any* number of matches)'.

The happenings with the *specific outcome*, that for which we want to know the probability, are:

'Finding a matchbox with 2 matches'.

Note: the *happening* could also be defined as '*Asking someone* for a matchbox...' (rather than 'finding... etc.'). This *will* affect the probability: You are more likely to find very *few* matches in a box that was left lying around.

Another example: The probability of getting through to anyone you know, having dialled a wrong number. '...having dialled the wrong number (with *whatever* consequences)' is the happening. 'Dialling the wrong number and getting through to anyone you know' is the *happening with the specific outcome group* for which we seek the probability.

Note: The apparently similar *happening* '...having dialled a wrong, *but existent* number' (i.e. being connected at all) leads to a very different probability: This is because not all numbers are allocated. The probability with the first *happening* would be smaller, because in many of the cases in which you dial a wrong number, you would not get anyone at all.

Another interesting distinction is that the *happening* of '...having dialled numbers at random' does not lead to the same probability as in the above cases, because the specification 'wrong number' excludes one of the people you know: 'wrong number' means that you *tried* one of the people you know, and having misdialled her, the scope of still getting someone you know is reduced by one. To clarify this point, think of a situation where you only *know* one person: When you dial randomly you stand *some* chance to chat, while if you mis-dial your only friend, you have to talk to yourself...

Note: the single '*specific outcome*' could be 'finding a box with 2 matches', or 'getting through to *a particular* person you know', while *specific outcome* ***group*** could be 'finding a matchbox with between 2 and 5 matches' (which consists of any of *four specific outcomes* 2,3,4,5), or 'dialling a wrong number and getting through to *any*one you know' ('*group'* consisting here of all the people you know).

Great care must be taken in describing fully and accurately the *situation*, or 'environment', in which the events take place.

In the example above, for instance, this might require adding "making phone calls in your home town", because clearly you would expect a somewhat smaller likelihood of the said *specific outcome* if you dial carelessly in Beijing. (Replace by *Mecca* in the Chinese edition).

In order to produce the number that will be associated with the probability of the happening with the *specific outcome (*or *outcome group)* we require:

1. The number of **happenings with the *specific outcome*** (or *specific outcome group)*

2. The number of *all* the distinct **happenings** (of the given description, and occurring in the given situation), **with *whatever* distinct outcome** (each of which is considered equally likely here).

The probability (of the happening with the specific outcome) will be the ratio of the two.

In the telephone example, the required probability would be worked out as:

The number of times you would be answered by a familiar voice (*having dialled a wrong number),*

divided by

The total number of times you dialled wrong numbers.

The above was described for probabilities as derived from actual happenings. As will be detailed later, probabilities can also be derived theoretically. In this case the definitions of the two numbers that need to be divided to get the probability, are:

1. The number of all the *distinct* ways in which (the happenings with) the *specific outcome (*or *outcome group)* can possibly occur

2. The number of all the ways in which (happenings with) *any possible* distinct outcomes can occur in the given situation.

Applying this *theoretical* way of deriving the probability to the telephone example (random diallings that get connected at all), the first figure (i.e. all the distinct possible ways of getting a familiar voice) is 'the total number of people you know there', and the second figure (all the possible distinct dialling-happenings with *whatever* outcome) is the town's population, assuming each person has his own (single) number.

By "distinct occurrences" we mean that they must be:

1. Independent of each other. (i.e. no occurrence must be liable to be influenced by another, as with lighting and thunder, for example).

2. Mutually exclusive (i.e. any one occurrence must not contain any of the other occurrences or parts thereof). For example, from occurrences defined as 'Dad, sister, nephew, parent', the 'parent' includes 'dad', (and so 'parent' should be replaced by 'Mum'). Another example: '4 or less' and '5 or more' are mutually exclusive, but '5 or less' and '4 or more' are not - each contains one member of the other.

A further condition is that in simple cases probability evaluations assume equal likelihood for each outcome (unlike 'being hit' which comprise 'hit by a car' and 'hit by a comet').

From here on, for the sake of brevity, we will use:

'**specific outcome**' for 'happenings with the specific outcome or specific outcome group' - the subject of the probability,

and '**happening**' for the 'happenings with whatever outcome', (and remember: *all* the happenings conform to the same well-defined description).

The probability (of the *specific outcome)* will then equal the number of cases of the *specific outcome* divided by the number of *happenings*. But remember always what these terms are the abbreviations of.

5.9 The implications of a stated probability

A probability of '1 in 10' for a *good news letter* does not mean, for instance, that if the nine last letters you opened all contained tax demands, the next one is bound to announce something really nice, such as your aunt and her dogs coming to stay for the summer. No probability other than 100% ever predicts the actual outcome of a specific happening. All that a probability such as that for the letters will tell you is that if you make many 're-runs' of the *happening* (i.e. receipt of letters) approximately one tenth of all letters will bring wonderful news. But you will never know in advance which one it will be.

There are three more points to note in this context.

1. The nice letters will not amount to *exactly* one tenth of your mail, but the more times you repeat the *happening*, the nearer to a tenth it will be.

For instance, if you receive 100 letters you can actually expect somewhere between 8 and 12 aunts, a 'spread' of 4. This figure, *relative to the 100,* is 4/100 = 0.04 .

If you survive to receive 1000 letters you might, similarly, expect between 92 and 108 nice ones, a 'spread' of 16, much greater than with 100 letters. But *relative to the 1000 received* this 'spread' is 16/1000 = 0.016, less than *half* the *relative* spread than with the 100 letters.

2. The foregoing is so providing that there is no 'organisation' of events (for instance your aunt colluding with the tax man...)

3. And that the situation does not change (for example the tax man becoming so offended by you ignoring his tax demands that he decides to forget about you).

What use can we make of the probability for a *specific* case, the next one, say?

One example would be making a choice between several equal-cost options. If each has a stated probability of success, these can be used for the purpose of *making a decision*, choosing the one with the highest probability. It is not *certain* that this would yield the best result, but what *is* certain is that some decision *has* to be made, and this provides the safest one. Thus when nothing is known about the other relative merits of a case, it is useful have a sensible aid for evaluations and decisions.

5.10 Determining the size of probabilities

There are two basic ways in which we can estimate probabilities:

1. Experimentally (i.e. by observing what happens, or has happened, in real or simulated situations. In other words, 'measuring' the probabilities.

2. Calculate them from knowledge of the situation.

Let us start with determining probabilities experimentally.

Firstly, there appears to be a contradiction: probability deals with uncertainty, but having done an experiment, we *know* the result for a fact. So, what we mean by *determining probabilities experimentally*, is using results from the experiments as *the best expectation* for the probability of future, uncertain, events. And as with the case of the letters above, we *expect* that the future events would diverge somewhat form the experimental results.

If we wanted the probability of scoring a hit on a dustbin from a distance of 100ft. we could simply throw stones until the neighbours complain, count the number of hits (the happening with the *specific* outcome) and divide it by the number of stones thrown (the happening with *whatever* outcome). (The *situation* is: *'from 100ft'*.)

However, the more interesting cases are those where, for some reason or other, we cannot carry out a 'test run' of the real thing. One reason could be that there is no time, because the outcome of the next available 'run' is precisely what we want to *predict*, so that we can *avoid* the real thing if that is what the prediction recommends. Another is where the location is inaccessible. Then, there are the situations where we have plenty of time and they could be on our doorstep, but we do not *want* them there or anywhere else, because they are too risky.

Consider an example of each of the above.

The value of the probability needs to be already known for the next forthcoming event, so this event cannot be used for experimenting:

> The first snow of winter is falling: what is the probability of tomorrow's train being delayed? (A certainty treated here as a probability for purely academic reasons)

Inaccessible location:

> What is the probability that lunar probes land in a crater from which the rover cannot get out?

"Too risky":

> Effectiveness of a new, untried drug.

Since the situations are unavailable for experiment, some form of 'simulation' of the situation is required. This might be data from real cases from the past, or a simulation that we design and run, such as a survey, or a physical or computerised model.

For the train we will use past experience, namely, past records from situations which we believe may be similar to the current one: same line, first snowfall etc., over the past five years.

For the lunar probe we will design a simulation: a model of the moon's terrain built by feeding into a computer thousands of numbers representing variations in the terrain, derived from photographs.

For the drug we have a physical model: a group of research-oriented mice.

The next stage is to identify the *specific outcomes* - those whose probability we seek, and the *happenings* - those with *all the possible* outcomes (*some* of which are the above *specific outcomes*). We then count the occurrences of the two in the historical, simulated, or test data. (The *ratio* of these two counts being regarded as the *probability* of the *specific outcome*).

In the case of the train, we have *happenings* of, say, 64 trains that travelled altogether on the line during the same period in the conditions stipulated above, and the 48 recorded delayed arrivals make up the *specific outcome group.*

The ratio of the two (i.e. cases of the *specific outcome / happenings)* works out at

48/64 = 0.75 = 75%. We may thus conclude that we have *approximately* 75% chances of spending the day building (PC) snowwomen instead of attending appointments.

(The exact reason for "approximately" is a long story, but it relates to the fact that even had all past 64 trains been delayed, i.e. a rate of 100%, you could still not be absolutely sure that *this* time the train might not, perhaps, skid there on time.)

There is more that can be done with this data.

So far, when we talked of probability, it related to the likelihood of one thing happening or not: the train being late or not being late; the probe lost or not lost; the drug safe or not safe.

There may, however, be cases in which we seek the probability of many other possibilities. In the case of the train, we might want to split the category of delayed trains into those delayed by less than ½ hour, between ½ and 1 hour, 1 to 1½ hours, etc.. As before, we could estimate these probabilities by counting the cases found in each range and divide by the total number of recorded trains (that ran in the defined situation). This is called the 'probability distribution' of the delays.

From this we could find the 'most likely' delay, and work out the (closely related, but distinct) *average amount* of the delay, e.g. 58 minutes, and also the 'spread' around the average, for instance whether most delays were concentrated between ½ and 1 hour, or whether they ranged more uniformly between a few minutes and the train never arriving at all...

Here again, we could not rely exactly on these figures for our next journey. However, the more cases we examine of similar cases in the past, the more accurate predictions will be. The more widely cases are spread around the average, the less useful the average will be. There is a particular way of presenting the spread in the most useful way; it is called "standard deviation". The dessert of this chapter will consist of showing how to work it out.

So, whenever you hear of an *average* you must pay attention to how the actual cases are spread around that average. The average between 1 and 79 is 40. If the winnings of an 'everyone wins' lottery are either £1 or £79 (with an even chance for each) you can say you get, on average, £40 (tickets, of course, costing £50...), but if the 1 and 79 refer to typical ages at which walking is not at polar-expedition standard, one would not say that 40 is the average age for walking difficulties.

In the case of the moon probe the *happenings* are 'simulated landings', i.e. a host of imaginary landings generated randomly within the range of the probe's known targeting accuracy, for instance, "falling anywhere within of 60 km of the intended target'. This is done by picking numbers between 0 and 60 at random (taken, for instance, from the minutes-past-the-hour of our train's *actual* arrival time...) and using these to determine how far from the target the probe fell in the simulation. Another random number, between 0 and 360 (deg.) determines the direction from the target. (This method results, correctly, in higher landing concentrations nearer the target.) For each landing location, some terrain data is known, and other needed details are, again randomly, picked from known ranges. This is used to determine where the probe would then roll to, and whether the rover could then get out.

This is repeated many times. Counting the proportion of randomly simulated landings that achieved no more than permanent (albeit free) parking of the rover (-the cases of the *specific outcome group*) gives an estimate of the probability that missions would fail because meteors got to the moon first and riddled its surface.

This simulation method of evaluating probabilities is called the "Monte Carlo method" (not that they make the best use of it down there...).

The mice are all given free samples of the drug (each mouse's 'free tasting session' is one of the *happenings*). Then one counts the ones that go on scurrying happily in the academic mazes and the ones that... don't (the sad *specific outcome*). The probability of the drug being unsafe is then the ratio of those that die to the total number of those that volunteered for the test (i.e. number of cases of the *specific outcome* / number of *happenings*).

5.11 But what if things change?

In all these cases the potential usability of the results depends on the assumption that the actual situation will not have changed to differ from the simulated one (or the one chosen from the past). In the case of the train we assume that snow, in northern winters, will continue to come as a complete surprise to the rail administrators. Regarding the applicability of the drug test, that humans have not meanwhile developed some significant distinction from mice. The moon presents a safer situation: it is too far away for anyone to mess things up.

As nothing really occurs without a cause, even situations which *appear* to be random are susceptible to change.

So how can we make use of the results of experiments / trials / surveys when we know, or at least suspect, that the situation changes?

We seem unable to get off this train: let us assume (or rather fantasise) that punctuality is improving year by year. In this case if we are to predict the prospects of delay tomorrow we do not consider the total number of delays in the past five years, but count the delays year by year, and for each year work out their proportion to the total number of trains that attempted the journey on the first day of snowfall. By comparing the results for each year, we look for an evolving pattern of change: do the figures change year by year by approximately the same amount (i.e. linearly)? Or do they change approximately by a fixed ratio? If we do find a pattern we continue (*'extrapolate'*) it into the next year. In this way we can look for a trend not only in the probability of delay, but also in the average time of delays.

The area of statistics that deals with extracting trends from historical data and evaluating their reliability is called *'regression'*. (Again, not a very good word: establishing trends is about looking back for the purpose of working out the future; 'regression' suggests going back and staying there...)

An entity that can assume different values is called a 'variable'. These include things like time, cost, success, depth, etc. ('house' is not a variable: it cannot be 'more house' or 'less house').

If repeated changes, all in the same direction (i.e. increasing or decreasing), in one variable of a situation are largely found to be associated with repeated changes, also all in the same direction, in another variable, we say that the two variables are 'correlated'. The measure of the strength of the connection is called the 'correlation coefficient'.

To give an example (not trains) we might take some variables related to the economics of your car: age (v1), annual depreciation (v2), annual cost of repairs (v3), number of accessory gadgets (v4). v1 is positively correlated with v3, and negatively with v2 (because the annual depreciation drops as age increases), but is not correlated in any way with v4 (you add gadgets, others steal them; there is no pattern).

5.12 Health warnings about statistics abuse

Three important points about correlation should be noted:

1. Correlation does not necessarily point to a causal effect! (The taller a man is the more often he gets invited to change light bulbs. Height is also an advantage in attracting the opposite sex. Therefore a correlation is likely to be found between number of lady-friends and the number of light bulbs a man has changed in his life. This does not mean that changing light bulbs gets one into bed.)

Where there *is* a causal connection between one variable and another it does not necessarily mean that it works the other way round. (Busy intellectuals often have no time to fuss about their appearance. One might therefore find a correlation between intellectual achievement and scruffiness, but this does not mean that letting your hair grow wild and tearing your jeans will win you a Nobel prize.)

2. Regressions are usually performed 'linearly', that is, it is assumed that when one variable repeatedly changes in equal steps, the associated variable also does so in equal steps. (It does not in most cases.) This assumption is often made without conviction,

simply because it makes the process of regression much easier, rather like losing a key in a dark field but searching for it in the street, because this is lit...

3. Although regressions are to do with change, they still assume some continuity. Going back to our train, if they get the Swiss to run the line, or close it down altogether, we can close the department of trains-statistics (but not forget how they work!).

It is also important to remember that the best use of such extrapolations is not to determine fate, but to forestall it (looking for cracks, not leaks).

When establishing probabilities experimentally, and this includes conducting surveys, attention should be given to the possibility of the tests or survey influencing the situation.

The most serious distortions occur in surveys. The seriousness lies not only in that surveys are often used to raise (our) money or support, but in that they can be rigged in a way that induces the interviewee to actively participate in the distortion.

The usual traps are:

1. Questions framed so as to lead to a preferred answer ("Will you vote for our goody party, which will increase benefits for the poor?" rather than "...party that will make *you* be the goody who pays for the wonderful benefits...")

2. Pre-set choices of answers that omit the ones they do not want (promoting of first class air travel: "Did you like the food? ...the service? ...the free tooth picks?" not "Do you really think the ticket should cost as much as a small aeroplane?")

3. An often overlooked problem with surveys is that interviewees tend to give the answers they think are nice to be heard saying. Much of what we *really* think and feel does not fall into this category...

When the result of a test gives a probability of 50% for an occurrence, or if all the possible events appear to be equally distributed, it does not (as you already know) mean that there is nothing guiding the events one way or the other. What it means is that the number of causes making things go left happen to roughly equal those making things go right.

There is, however, another meaning to *50% probability*: when probabilities cannot be estimated by observation they must be produced by calculations based on what we believe we know about the situation. If we know absolutely *nothing* about why outcomes should turn out one way or another, what are we to say? It is best, really, to say *nothing*, but then there are times when we are compelled to say *something*.

In such situations of total ignorance about the (say, two) possible outcomes we attribute them a *probability of 50%*, not meaning that we *know* the opposing forces to be equal to each other, but to state that we have no grounds for declaring either of them a winner.

In either of the above situations, namely, where we know of the 'equal opportunities' or where we know nothing at all, what *actually* happens we call '*randomness*'.

5.13 Evaluation of probabilities by calculation

Let us begin with the above random situations, because they are simple and common.

In a situation where the *happenings* have only 3 possible different outcomes and all are equally likely to occur, the chance for each of them is "one in three", i.e. 1/3, or one third.

If, instead of 3, there are 30 different possible outcomes, all having an equal chance of occurring, the probability that any one of them will happen is much smaller: one thirtieth (1/30).

Thus we can generalise by stating that the probability of any particular outcome (i.e. the *happening* with the *specific outcome*) is

1 / total number of possible *happenings,* each with a different *outcome*),

or, in short:

1 / total number of possible different *outcomes*).

Let us call the occurrence of the *specific outcome* a "win", and the probability of its occurrence the "likelihood of winning". Now let us suppose that in addition to the one specific outcome that makes you a winner, there is another i.e. you win a prize if any of these *two* separate outcomes occur. Since both outcomes have equal likelihood of occurring, you regard prospects as doubled: given 30 possible '*happenings (with different outcomes)*', your likelihood of winning are now

$$2 \times 1/30 = 2/30 = 1/15.$$

If, out of the total of 30 *happenings* there are 20 different outcomes that are deemed 'winners' your likelihood would rise to

$$20 \times 1/30 = 20/30 = 2/3.$$

(With odds that high the prize is usually a promotional leaflet telling you how to *lose your* money...)

This concurs with our definition of (a calculated) probability as being the ratio between

- the number of all possible different ways in which the *specific outcome group* can occur, and

- the number of all possible ways in which the *happening* (with *whatever* outcome) can occur.

Again, it is assumed that all the possible individual outcomes are equally likely.

For the experimental determination of probabilities we counted these two numbers as they actually occurred, but here we *calculate* what we *think* they should be, basing ourselves on what knowledge we have of the situation.

It should be clear by now that no evaluation of probability is possible unless we *start* by understanding exactly how the *specific outcome* is defined.

Anyone who grasps this will cease to be mesmerised by the ninety-seven incorporated forms of paranonsenseology (examples of which will be given later).

With that in mind, let us work out some probabilities.

Assume that the first snowfall of the year brings everything to a standstill. What is the probability of this being good news, namely, that it falls on a weekday and creates an extra day off?

The snow does not know or care what day it is, so the question translates into: what is the probability of a randomly chosen day being a weekday.

The *happening* here is '(any) day', of which there can be 7, and the *special outcome group*, the probability of which we seek, is 'weekday', of which there can be 5. Therefore the required probability is 5/7 (= 0.71 = 71%).

Prime numbers are ones which cannot result from multiplying integers (i.e. whole numbers, not fractions). This means that there is no integer (other than 1 and the prime number itself) into which a prime number can be divided, with the result, too, being an integer. 15 is not a prime number because it can be reached by 3 x 5 (and can, thus, be *divided* by 3 and 5). 59 *is* prime (try!).

What is the probability that a number drawn at random between 1 and 25 will be a prime number? (The *happening*: drawing a number between 1 and 25; the *specific outcome group*: 'prime numbers')

The prime numbers between 1 and 25 are 1*, 2, 3, 5, 7, 11, 13, 17, 19, 23: ten in all. 10 out of the total number of possible draws is 10 / 25 = 40%.

Now if a number between 3 and 25 is chosen at random, what is the probability that both it and the next number will be prime numbers?

Zero. It cannot happen. Because of any two consecutive numbers, one must be even, i.e. divisible by 2, and therefore not a prime number (unless of course it is 2 itself; this was the reason for excluding 1 and 2 from the question). This was a little example of mathematical proof.

These examples were chosen in order to avoid more dice stories...

What is the probability of *getting* a total of 3 or less when throwing two dice?

In this case, our knowledge of the situation is that all dice are perfectly unbiased when they are used in examples of probability. Let us suppose that only one of the two dice is a cube, and the other is a tetrahedron - a 'cut-price' pyramid, namely: a pyramid with a triangular base, and so only three (triangular) faces rising to the apex, i.e. a total of four identical faces numbered 1 to 4.

---

* the conventional definition of *prime numbers* does not include '1' but logically it should, and we will do so here

This de-banalization of dice stories compels us to refer to 'face down' outcomes, as 'face up' in the case of the tetrahedron might be a little confusing (like "meet me next to the Pentagon"...)

Let us begin by counting the number of all the possible outcomes when these two dice come to rest - the total of the *happenings*. In the following series the first number in each pair is the one under the tetrahedron, the second number the one under the cube: 1&1, 1&2, 1&3, 1&4, 1&5, 1&6; 2&1, 2&2, 2&3 etc.. What is etc.? For each of the 4 possible numbers under the tetrahedron the cube can sit on any one of its six numbers, therefore the total number of combinations of the cube and tetrahedron bases is 4 x 6 = 24.

Next we must establish which, and hence how many, of these combinations comply with the specification of the *specific outcome group*: getting a sum of 3 or less.

A sum of 1 is impossible with two dice; a sum of two occurs when both dice rest on their '1'; but a sum of 3 can occur in two ways: with the tetrahedron on 1 and the cube on 2, and vice-versa. The total is 3.

The probability of getting a sum of 3 or less is thus 3/24 = 1/8 = 12.5%.

This business of "how many different ways..." is the backbone of evaluating probabilities.

In some cases we just have to try and think of all the possibilities that are distinct cases of the specified outcome. For example, consider a registered letter that has to be collected from the post office. What is the probability that it can wait, i.e. that it is a bill? First you need to count the creditors whom you ignored so often as to expect a registered final notice (the 'cases' which make up the *specific outcome group)*, then count *all* the people you can think of, (including the above creditors), who could possibly wish to contact you so badly that they would queue at a post office and pay for registering a letter (the *complete* set of relevant *happenings* - those with *whatever* outcome).

Other situations may be more ordered, so we can calculate the number or possibilities instead of counting them, and, what is more important, know that none have been omitted.

Often, outcomes are 'composed' of the values of several *elements*. Two, in the above dice case.

When enumerating or calculating the number of possible ways in which a specific outcome can come about it is very important to note how, exactly, the composition of the elements is defined for that outcome:

Firstly, it obviously matters how many elements the outcome is composed of.

Then it matters how many different values such elements can assume. (In our example above, four under one die, and six under the other. In simpler examples, though, these numbers would be equal.)

In the most common cases it is assumed that each of the various values of the elements is equally likely to occur. (When this is not the case, their *relative* expected occurrence can be incorporated by a process called 'weighting' but we will not bother about this here.)

Further important distinctions exist and must be specified: in some cases it only matters *which* values of the elements make up a given outcome, but in other cases it also matters in what order they come, i.e. cases the same set of values of the elements but in a different order are regarded as different outcomes. For example, if the elements were playing cards, and the outcome were a *hand* of 2, 5, 4, K, then K, 4, 2, 5 would count as the same outcome. But if the outcomes were defined as *words* of three elements (letters), and a specific outcome defined as being composed of *g, d & o,* then naturally, 'god' and 'dog' would count as different outcomes, certainly from the cat's point of view.

And, it matters a lot whether or not any one value can be used more than once in a composition: For example, when the elements are the two participants of boxing matches, and one wanted to enumerate all the possible different 'compositions' that could be made up from 20 men who lack other stimulation. As healthy people do not wrestle with themselves, the values of the elements in any pair cannot be the same.

Note: the order, here, of the values of the elements does not matter, because Jim fighting Jack is the same as Jack fighting Jim. However, were Jack and Jim more civilised and played chess, with the first of the pair playing white and the second black (if you play well the colour you play matters), Jim v Jack and Jack v Jim would then count as different events.

The probability of an outcome can thus be defined as the ratio of:

the number of the (distinct, allowable) 'compositions' as defined to make up the *specific* outcome,

to

the number of *all possible* (distinct, allowable) 'compositions'

('distinct, allowable' in accordance with the above detailed criteria).

As an example, let us work out the probability that a monkey will come up with the word 'banana' by hitting letter-keys on a keyboard at random six times.

The elements are the six letters, each having 26 possible values. There is only one combination of these that produces the specific outcome, the word 'banana'.

We must therefore divide this one combination by the number of all possible distinct 'words' of six letters that can be made up by combinations of the 26 letters.

(Letters may be repeated in a word; the order of any given set of letters matters, as we noted in the above divine canine example; the 'words' do not have to make sense, they only have to differ from each other.)

To show how the number of all the possible six letter words can be calculated, let us first imagine an alphabet of only 4 letters: *a, b, c, d.* We will call the position on the left of the word the 1st position; thus in *bacd* *b* is in first position, *a* is in second position etc..

Let us start with *a* in the first position; we can pair it with four different letters in the second position: *aa, ab, ac, ad.* We can do the same with *b* in the first position:

ba, bb, bc, bd, and then with *c* in the first position: *ca, cb, cc, cd*, and finally *da, db, dc, dd.*

We had 4 (the length of our economy alphabet) possibilities for the first position, and teamed each of these with the 4 possibilities in the second position, giving us 4 x 4 = 16 combinations. Now let us take each of these pairs *in turn* and make them into a triplet by adding a letter in the third position. We begin with the first pair, aa, giving it, one at a time, each of the (4) possible neighbours in the third position: *aa a, aa b, aa c, aa d.* Then we do the same with all 16 pairs, giving each pair each of the 4 different partners in the third position, which gives us 4 x 16 different triplets, or 4 x 4 x 4 (4^3). Finally we repeat the process with these (4^3) triplets each getting, in turn, four different neighbours in the fourth position, which gives us a total of 4 x 4x4x4 different quartets (4^4).

The same process with a word of 5 letters would give 4 x 4x4x4x4 (4^5).

A word n letters long would give 4^n.

Had we 7 letters to choose from we would have 7 x 7 pairs, 7 x 7 x 7 triplets, 7^5 words of length 5, and 7^n words of length *n*.

Were our alphabet to consist of L letters we could make up L x L pairs, L x L x L triplets, *and* L^n *n*-length (mainly nonsense) words, i.e. L^n distinct sequences of *n* length.

The English alphabet of 26 letters allows a total of 26^6, (300 million!) different 6-letter words. Only one of these, poor monkey, is 'banana'.

At one keystroke a second, i.e. ten sextets a minute he would need sixty years to type 300,000,000 trials. So, is he likely to issue a typed order for a banana in his lifetime? No way. Because he would also need to keep checking a rather long record to make sure he does not *repeat* any word... (And so, monkeys decided to reach bureaucracy by evolution rather than by trial and error.)

Given a menu of 10 symbols, the number of different sequences of 3 symbols is 10^3. Supposing that the first of these symbols were $ and the last one £ the list of all triplets set out in an orderly way (which is the *only* way to ensure its completeness) will begin with $$$ and end with £££. If the first symbol is not $, but 0, and the last one is 9, the list starts with 000 and ends with 999, and is 10^3= 1000-long (note: the 000 is included in the count, that is why 999 is the 1000th member.)

We have already seen the reason for which we write numbers in the way we do. The progression of a thousand steps from 0 (or 000) to 999 provides a reason from another point of view: it is simply the orderly way of recording all the 3-long sequences that can be made up of the ten available symbols 0-9. In the example given above the list started with aaa, aab, aac, which corresponds to 000, 001, 002, etc..

Now let us look at a case in which no repetition is allowed, in other words in which an element can only appear once in any one sequence.

Imagine a day-tour operator in a resort with only 5 tourist attractions. Only 3 sites can be visited in any one day. The tour operator wants to see how thick a brochure he can compile, in other words how many different 3-site tours he can offer. He therefore needs to work out how many different sequences of 3 he can make up out of the 5 sites. The necessity for no repetition arises from the unlikelihood that visiting the same site twice, not to mention three times on the same day will go unnoticed.

Having rushed his tourists through breakfast, he can take them to any one of the 5 sites. For each of the 5 possible options for the morning excursion, he has a choice of 4 other sites to take them to in the afternoon instead of allowing them the rest they have yearned for all year, making 5 x 4 possible two-site tours. For each of these 5 x 4 possibilities, he still has a choice of 3 unvisited sites with which to finish them off (and miss the free buffet dinner at the hotel). The tour operator thus has the option of 5 x 4 x 3 different 'compositions' of sites on the same day, any of which will be a different way for the dream of a rest remaining a dream.

Here is the full list of the site-compositions: (for the colour pictures click on...) The sites are abbreviated *a, b, c, d, e*. The key to ensuring that the list of 'compositions' is complete is to display them in an orderly way. Ideally the 'trios' should be in a single long column; to save paper we will set them out in five parallel columns. We will list all the trios beginning with *a* in the first column, those beginning with *b* in the second column etc..

In the second position of each trio we put one of each of the letters from *a* to *e*, in ascending order, but omit the letter already used in the first position. For instance, in the column of trios starting with *c* the sequence of letters in the second position will be *a*, then *b, d, e*. However, before incrementing the letter in this second position we fill the third position with, one by one, the letters which we have not yet used in the positions to its left. For instance, in the 4th column, with *d* in the 1st position, there are 3 rows with *b* in the 2nd position: one ending with *a* in the 3rd position, one with *c*, and one with *e*.

In this way we display the choice of 5 x 4 x 3 different tours offered by our tour operator:

a b c <	b a c <	c a b <	d a b	e a b
a b d	b a d	c a d	d a c	> e a c
a b e	b a e	> c a e	d a e	e a e
a c b <	b c a <	c b a <	d b a	e b a
a c d	b c d	c b d	d b c	e b c
> a c e	b c e	c b e	d b e	e b d
a d b	b d a	c d a	d c a	> e c a
a d c	b c d	c d b	d c b	e c b
a d e	b d e	c d e	d c e	e c d
a e b	b e a	> c e a	d e a	e d a
> a e c	b e c	c e b	d e b	e d b
a e d	b e d	c e d	d e c	e d c

In total: 5 x 4 x 3 = 60 'programs'

Had the tour operator had access to 8 sites and designed tours visiting 5 sites in one day (the tourists now yearning to return to work) he would

Choose as the 1st site one of the available 8 sites. Then for each of these
choose as the 2nd site one of the remaining 7 sites, then for each of these pairs
" " " 3rd " " " " " 6 ", " " " " " triplets
" " " 4th " " " " " 5 ", " " " " " quartets
" " " 5th " " " " " 4 ", thus completing the quintets

The total number of different tours is 8 x 7 x 6 x 5 x 4 = 6,720.

Which is too many. So let us return to the 60 different possible ways of covering 5 sites at the rate of 3 a day. It might occur to our tour operator that his grateful clients do not care which item on the programme disrupts which part of the day as long as they see all the ruins, in other words, that the order of visits does not matter. In this case, tour *b,a,d* would be tour regarded as equally good as *d,a,b*.

A look at the painstakingly compiled table above will show that every possible composition of letters reappears in different orders. For instance all the tours including sites *a b c* are marked * and those including *a c e* are marked >.

How many different orders are there for any given 3-letter set?

Again, a similar story: say, the set is *x y z*, we seek an orderly way of listing all the possible orders. As the first one, we have 3 to choose from (***x, y, z***); for each of these starters we are left with 2 choices for a partner: ***xz, xy, yx, yz, zx, zy***; and for each of these (3 x 2) pairs, to find them a third member there is not much of a choice - only one is left (for the *xz* there is only the *y*, for the *xy* there is only the *z*, etc.).

So, there are 3 x 2 (x 1) different possible orders for a combination of 3 items

(***xy*** *z*, ***xz*** *y*, ***yx*** *z*, ***yz*** *x*, ***zx*** *y*, ***zy*** *x*).

For the number of all possible different orders of 4 items the story goes: we have 4 choices for the first; for each of these 3 possibilities remain for making up a pair, giving 4 x 3 different pairs; for each of these, 2 possibilities remain for making triplets; and for each of these 4 x 3 x 2 triplets, as to the 4th member there is only one available (this is what 'x 1' stands for at the end of 4 x 3 x 2 x 1).

In our tour operator's brochure of 60 tours every distinct tour of 3 sites appears in 6 (3 x 2 x 1) different orders. This means that there are in fact only 60/6 = 10 tours that are distinct in the sense that there are no two tours that have all three sites in common:

abc, abd, abe, acd, ace, ade bcd, bce bde cde

Notice the order of these ten triplets. They are the result of 5x4x3 / 3x2x1 (=10).

(The method for this listing is "ascend only": here, contrary to the table/listing in the previous page, letters along each triplet appear only in ascending order.)

In the case of the 8 sites / 5-item tours, the number of distinct combinations is

8x7x6x5x4 / 5x4x3x2x1 = 56.

5.14 Combinations and permutations

So far we have not used precise *terms* to distinguish the two kinds of enumeration of the 'compositions' we have discussed: the one in which the *order* of a given set of items matters, and the one in which it does not.

Where we only count distinct contents of the compositions, i.e. where no two 'compositions' contain exactly the same set of items, irrespective of their order, we are counting *combinations*.

Where we 'inflate' the list of combinations to distinguish between all the different possible orders in which the items of a given combination can appear, we are counting *permutations*.

Note that in neither case can items be repeated within any one 'composition'.

Thus counting all possible 4-long compositions of elements, chosen from a list of 9 elements:

- the number of *permutations* is 9 x 8 x 7 x 6 = 3024 and

- the number of *combinations* is 9 x 8 x 7 x 6 / 4 x 3 x 2 x 1 = 126.

See the order in which these two are calculated: combinations cannot be worked out directly; the *permutations* must be worked out first, and this total is then divided by the number of different possible orders of the elements in each combination. (Remember: it naturally follows that there are always fewer combinations than permutations.)

Naturally, there is a shorthand way to write 7 x 6 x 5 x 4 x 3 x 2 x 1, and, of course, it is totally nonsensical: the written abbreviation, would you believe it, is 7! and it is called '*factorial*'.

Why, Why?

Surely, 7↓ would be more suggestive, and calling it '7 *multdown*' would tell it *all*...

Earlier, when we considered the enumeration of all 3-digit numbers that can me made up from 10 symbols (1000), we allowed repetition of digits. Let us now look at a digits case where repetition is excluded.

People are often surprised by the large number of cars which have a repeated digit in the 3-digit part of the number plate. But what *is* the probability of the next car you see having such a number plate?

Counting all the different ways this 'event' might occur would be a messy business: the repeated digits might be in positions 1&2, 1&3, 2&3, or in all three positions. There is a simpler way of doing this which amounts to a neat trick.

We said earlier that the probability of something occurring is represented by one of two fractions which make up 1. The other fraction represents the probability of that thing *not* occurring, Why? because it is always *certain* (i.e. probability =1) that anything will either occur or not occur. In this case the "not occurring" means all three digits being different.

This is the probability which we will work out, because it is much easier to do so. All we need do then is subtract it from 1 (to get the probability of "repeated digits").

The probability of all three digits being different is:

> the number of ways in which the *special outcome* ('3 digit number with no repeated digit, order matters') may occur, i.e. the number of *permutations* of three digits out of ten
>
> divided by
>
> the enumeration of *all* the '*happenings*' (all possible 3 digit numbers) that may occur, i.e. 1000.

We now know how to calculate the *permutations*: choose any of the ten available digits as the first of our set of 3 digits; this can be paired with any of the remaining nine, giving 10 x 9; finally, this sum can be linked to any of the remaining 8 digits to complete the triplet: 10 x 9 x 8 = 720.

Dividing 720 by the 1000 possible 3-digit numbers gives 0.72, or 72%. This is the proportion of license plates expected to have no digits repeated, therefore the probability of plates *with* repeated digits is 28%.

28% lies between one in three (1/3 = 33%) and one in four (1/4 = 25%). With a proportion like that it would be surprising if we did *not* see licence plates with repeated digits all over the place.

If number plates had 4-digit, nearly half of the cars would have a repeated digit:

Permutations of 4 out of 10 (10 x 9 x 8 x 7 = 5040) divided by the *total* number of 4-digit numbers (10^4) = 0.504, i.e. the fraction with no repeats is 0.504. Therefore the fraction with repeated digits is 1 - 0.504 = 0.496, or nearly 50%.

5.15 Beware of MULTDOWNS

In Chapter 3 we saw what large *Num.mults* can do to you (i.e. a^n with a large *n*). That is nothing compared with the havoc large MULTDOWNS can create.

Here comes the famous *travelling salesman* story: he wants to find the fastest way to visit all fifty state capitals of the United States and return home. (Note: he is not seeking the shortest distance but shortest *time*.)

This does of course depend on the order in which he visits the cities.

First, having considered all possible means of transport, schedules, connections etc., he will have to determine the shortest possible time of completion for each one of the possible sequences (orders) in which the cities can be visited.

(Finding the shortest time for just one of the sequences is a major task in itself; my travel agent spends half a day doing this for a sequence of only four cities.)

So he will have to use a computer. Not the one you have, not the one you *want*, but the one your *kids* want. One that could calculate the shortest time for one whole 50-city sequence in one billionth of a second. And since it will have to do this for all the different orders of 50, (and then find the fastest one), the computer would benefit from a few more computers sharing the task. In fact, the globe's entire surface, land and sea, covered with computers, four to a square metre in stacks ten high. Roughly how long will it take them to do the job?

Forty five million million million million million years, give or take an hour. Why?

Here we go again: as the first step he has 50 cities to chose from. For each of the 50, there remain 49 to choose from as the second step. So far, for only the first two steps we have 50 x 49 different possible sequences; For each of these there are then 48 different choices for the third step, i.e. 50 x 49 x 48, or 117,600 different possibilities for only the first three steps, etc. for all the 50 steps. But this is BIG 'etc.': 50↓, or

50 *MULTDOWN* ('factorial'):

50 x 49 x 48 x 47 x 46 x 45 x 44 x 43................ x 20 x 19 .... x 5 x 4 x 3 x 2 x 1

No one messes with numbers like this.

To estimate this monster's size let us, instead of multiplying 50 numbers ranging from 1 to 50, take one number *N* and repeatedly multiply it 50 times, i.e. *Selfmult* N, *Num.mults* 50.

This number N would, of course, have to be something between 1 & 50, in fact- just under 20, to give the same result, i.e. 50↓ = 20^{50} (approx.). How much is this?

$$20^{50} = (2\text{x}10)^{50} = \underbrace{2\text{x}10 \text{ x } 2\text{x}10 \text{ x } 2\text{x}10...}_{50} = \underbrace{2 \text{ x } 2 \text{ x } 2 \text{ x}...}_{50} \text{ x } \underbrace{10 \text{ x } 10 \text{ x } 10 \text{ x}...}_{50} = 2^{50}\text{x}10^{50}.$$

The 2^{50} alone is already a million billion, and this little starter then acquires 50 zeros with the compliments of 10^{50}. Big*.

How much is 1000↓ / 998↓?

Working out 1000↓ entails multiplying 1000 mostly 3 digit numbers together; jotted down the result would be longer than your street... and the figure would be almost as long again for the 998. Then, dividing two kilometre-long numbers makes 'long division' something of an understatement...

If we just contemplate the sum for a moment and think what goes on there, we will see that doing it as described above would be (literally) as pointless as "to hell and back".

What is 6↓ / 5↓?

$$\frac{6 \text{ x } 5 \text{ x } 4 \text{ x } 3 \text{ x } 2 \text{ x } 1}{5 \text{ x } 4 \text{ x } 3 \text{ x } 2 \text{ x } 1}$$

The 5 x 4 x 3 x 2 x 1 on top is divided by the same on the bottom, leaving just the 6.

In the case of 1000↓ / 998↓ only the 1000 x 999 on the top survive; the rest (x 998 x 997 x...etc., or 998↓) are mirrored on the bottom, so the two sides kill each other off down to the last man, leaving only a graveyard of 998 'x1's... and of course the 1000 x 999 = 999,000.

---

* There are methods to short cut this search through all the possibilities. They belong to an area of mathematics called Operations Research.

At the beginning of the twentieth century composers ran out of real musical ideas, so one of them contrived a 'method' for constructing musical themes, which consists of using all 12 notes of the scale, but using each one only once. The problem with this method is that it provides $11^{\downarrow}$, or forty million different ways of tormenting the human ear (*real* art cannot be 'produced' from recipes).

This figure is $11^{\downarrow}$ and not $12^{\downarrow}$ because there is, in effect, only one first note: were all 12 starting points used, each 'melody' would reappear 12 times here and there among the $12^{\downarrow}$ sequences, only transposed ('shifted') to various starting points (but sounding just as bad). So, the net number is only $12^{\downarrow} / 12 = 11^{\downarrow}$.

We have seen how the probability of an event depends on the number of distinct ways in which the specifically defined outcome can occur in a given situation. Equivalently, the probability increases with the number of times we are allowed to repeat the try.

5.16 Some great enlightenments, and be fooled no more!

We have already seen that our intuition is not at its best when estimating probabilities.

There are two main sources for the mis-appraisal of probabilities (and belief in miracles...)

1. The first is inattention to 'non events'. A resounding success, now and then, among many 'failures' does not appear so miraculous. However there is a natural tendency to ignore the many failures that envelope the conspicuous success. The successes thus appear more frequent than they really are.

When someone phones you just as you were thinking of her, your amazement at the coincidence stems from your inattention to the many times when people have *not* called just as you were thinking of them (*any* of them, not just the one who *did* eventually call).

You might try an experiment: point at the telephone once every minute all day long. Nobody will pay much attention. However when (not *if*) within two seconds of your command the telephone does ring, everyone, including, and this is the problem, yourself, is liable to invest you with remarkable tele(phony)pathic powers. The 'ringless pointings' and the 'pointingless rings' will have gone unnoticed. The two-second coincidence (which, incidentally, is bound to occur in one out of about thirty calls) can set you up as a cult leader.

2. The other cause of intuitive mis-appraisal is far more serious. It is failure to notice that the description of a predicted *specific outcome* is not limited to a single outcome, but encompasses a very great number of different possible outcomes. Each one of these may indeed be very improbable, but the probability of any one of them occurring, in other words the probability of the sum of all these minor but vastly numerous probabilities, is not small at all, and therefore not at all surprising. (In our terminology: what appears as a *specific outcome* is really a *specific outcome* ***group***).

This inattention to lack of definition of an event is what fortune-tellers depend on. Fortune tellers will first condition their subject's credulity with impressive revelations about his past, or rather, about pretty well anybody's past: things so general than anybody can recall details of their past that fit them. What the subject overlooks is that *his* recollection of these details does not mean that the eyes across the table gazing into the crystal ball know anything about it. It is he, the subject, who has unconsciously added the specificity to the general statement, and thereby blinded himself to its unremarkable generality. Now that he is conditioned come the 'wide angle' view of the future: he will 'find great love', but never a little help such as a phone number…; a 'long journey' might mean anything from finding yourself finding yourself in Nepal to an average bus journey in London…

Consider a business venture with a 'guaranteed 90% success râte'. It sounds promising, but what does 'success' refer to? Making any money at all? Making a little more than the paltry interest offered by any bank? Making enough to consider tax exile in Switzerland? And then there is the question of *when*? Success within one year, or within the life time of your great grand-children? Unless these things are specified, it really does not matter whether the promised chance of success is 90% or 71%. The wider the range of what counts as the specific outcome, the greater its probability.

This example is perhaps rather transparently 'fuzzy'. Let us look at a case where the outcome *does* appear to be narrowly defined, but *still* is not: the 'same birthday' trick. (The method of calculation to be used here is not quite correct, but it makes the intended point clearer. The correct way will be indicated later.)

If I were to walk you through a series of train compartments, each occupied by twenty passengers, and from time to time announce "I sense that two people in the next compartment have the same birthday", and, upon checking the passengers birthdays, was found to be right half the time, you might be very impressed. You need not be, even your investment adviser could get *this* prediction right.

You would be impressed because you saw the chance of such a coincidence as being 1 in 365.

Your attention is restricted to *one particular* case - to those two people who were *found* to share the same birthday (indeed, for *those two* the probability *was* 1/365). But I, like all high achieving prophets, did not talk about those two specifically. What I said was "...two people have...", meaning *any* two. The point is that *any* in this case means a *lot* of 'any'; given 20 people, there are many pair-candidates for a common birthday:

For each of the 20 passengers there are 19 possible others with whom to form a pair, each of which has a common-birthday probability of 1/365. This yields 20x19 possibilities of distinct pairs, but the sum must be divided by 2, because every pair is matched twice: consider Sue and Mo. In one case Sue appears as the first member of the pair - one out of the 20, and Mo is taken from the remaining 19 as the second member of the pair. In the second case Mo appears as the first member of the pair - one out of the 20, and Sue is taken from the remaining 19 as the second member of the pair. But here the order of the people in the pair does not matter, so the *distinct* chances are 20 x 19 /2 = 190, each with a probability of success of 1/365 (like buying 190 lottery tickets which each have one chance in 365 of winning). In all, then, the probability of my being right was

$$190 \times 1/365 = 0.52 = 52\%.$$

A 'same birthday' *is* a rare occurrence, but this rarity is given many chances.

The effect can be illustrated in a multitude of other ways. Imagine you board a bus in a town you have never been to before and sit next to a stranger. The next day, in a different town, the same stranger stands before you in a queue. That really would be unlikely, a chance perhaps of one in a million, if that were the population of the region. But then try to imagine all the many other amazing coincidences of this kind that could occur... The list is endless, so one thing or other of this kind is *bound* to cross your path now and then.

Were you to get up one morning and think of (or if I would predict) a very specific and highly unlikely event, and precisely that event befell that same day, that would be very remarkable indeed. But when such things *seem* to happen, the actual description of the prophecy conceals a great scope for an innumerable variety of naturally occurring rare events, or for a multitude of different ways for the event to occur, as in the case of the birthdays.

As noted above there was a small inaccuracy in the calculation of birthday probabilities. In this case it is not quite correct to add together the chances provided by the different pairs. Again, the situation can be clarified with dice: If you throw a die once, the chance of a 3 tuning up is 1/6. If you throw it twice, the chance is apparently doubled, but that is not quite so; if it were, the chance of a 3 when you threw 6 times would be 6 x 1/6 = 1, i.e. certainty. In reality of course, you might cast dice until they wore holes through the table and still not be *absolutely* certain of throwing a 3.

The rule is that chances can only be accumulated for 'mutually exclusive' events, i.e. where the occurrence of one event precludes occurrence of the other. This is not the case with dice: casting a 3 in one throw does not preclude the possibility of 3 turning up on the second throw too; therefore the probabilities of the two throws cannot be simply added up.

An example where summing *non*-exclusive things would lead to nonsense is the consultant who invoices 7 different clients, each for 6 hours' work, all done on the same day... Either those services cannot have been exactly 'mutually exclusive', or the consultant's day can not have too many hours left to spend all that money.

The simplest, and yet correct way of working out the combined probability of several non-exclusive trials, such as the birthday trick, is to proceed the way we did with the repeated digits on car number-plates: first work out the probability that *none* of the passengers have a birthday in common, and then subtract the outcome from 1. Doing it this way the (now accurate) result is that for a success rate of 50% you need 23 people, not 20, but this is still good enough to amaze many who have not yet read this book.

In this context, here is an interesting distinction (already hinted to above).

As we know, when casting a die the probability for any particular number face-up, 3 say, is 1/6.

There are two, *apparently* equally effective ways of increasing our chances of a 'win'. If we wanted to double our chances, we might either:

a. cast the die twice instead of once

b. cast the die once, but allow a 'win' if either of *two*, rather than just one, pre-determined numbers turn up, say 3 *or* 5).

We have seen that in the first case *a*, our chances of winning do increase, but are not quite *doubled*, because casting a 3 on the first attempt does not preclude the (small) possibility of also casting a 3 the second time, and vice-versa, i.e. they are not 'mutually exclusive'.

We may visualise why this undercuts doubling the chances of winning this way: we may regard the chance of a 'win' as an allotment of luck ('luck' being the outcome of a 3 when the die is cast). In the rare outcome of 3 facing up in *both* throws (one die cast twice being the same as two dice cast once), it still counts only as a single win, so a portion of luck (outcome of a 3) was 'wasted' in the course of repeated trials as a 'doubled turn-up'.

In case *b* the winning casts are mutually exclusive. In a single cast the outcome of 3 precludes the outcome of the other winner, 5, and vice versa. There is no possibility of one stroke of luck being accompanied by another, unusable, stroke of luck. Therefore, by allowing two, alternative winners, the probability of winning is exactly doubled. *Moral*: better one shot at a big target than many shots at a small one. (Fact: there are more surviving ants than elephants...)

Case *b* showed us that when we know the probability of various, mutually exclusive, outcomes we obtain the probability of 'one *or* the other' of the outcomes by *adding up* their individual probabilities. (In the case of the dice, we added the 1/6 probabilities of the 3 and of the 5.) Now let us see how we may combine the probabilities of various outcomes, this time to find the probability of *all* of them occurring (i.e. one *and* the other as opposed to one *or* the other).

Before we begin, let us consider what we might expect. We found that our chances of winning are greater when they depend on the occurrence of *any of several* outcomes rather than on the occurrence of only a *single* outcome. However, if the occurrence of just *one* specific outcome is not enough to get us the prize - if we get it only if *two* outcomes occur (e.g. not only pretty, but clever too), clearly we expect the chances of winning to be *diminished.*

'Meeting' implies being in the same place at the same time (two outcomes, *both* of which must occur).

Once a week, on a day chosen at random, someone boards a daily train at its starting point. The train has five compartments. To meet that person by boarding the train further down the line, you must board it on the right day *and* get into the right compartment.

The probability of striking the right day of the week is 1/7. This means that if you boarded the train weekly for 700 weeks that person would be on the train on approximately 100 occasions. But this alone will not make you meet: your (randomly) chosen compartment, too, has to be the right one, the probability of which is 1/5. So only in 20 out of the 100 cases where the other person is actually on the train, you *will* meet. 20 out of the total 700 trials that you made is 1 chance in 35.

How is this 1/35 derived from the separate probabilities of 1/7 & 1/5? The chance of boarding the train on the only day the other person is on the train was 1 in 7 (1/7); it is only on one fifth (1/5) *of* this 1/7 of the occasions that you enter the right compartment. We know that '*of*' means *times* in arithmetic ('two ***of*** something' = 2 **x** something); therefore a meeting will occur on approx. 1/5 x 1/7 = 1/35 of your 'trials'. 1/35 is, of course, much smaller that each of the 1/7 and the 1/5, so, as we expected, the probability of *meeting* is much smaller than either of the separate probabilities - that of the right *time* (the day) and that of the right *place* (the compartment).

The rule, therefore, is that where there are two (independent) outcomes, with probabilities P_1 & P_2 the probability of both outcomes occurring together is $P_1 \times P_2$.

If *some* (any) comet passes within view of the earth on average once every 400 nights, i.e. the nightly 'comet probability' 1/400, but you try to watch them from London, where only one night in 20 is sufficiently clear to distinguish comets from hallucinations in the yellowish mist, your probability of enjoying comets is 1/400 x 1/20 = 1/8000, or once every 22 years.

This rule only holds, however, when the outcomes are independent of each other, i.e. when the occurrence or non-occurrence of one has nothing to do with what the other chooses to do. This is certainly so in the case of comets, whose travel plans show a total disregard for English weather. In the train scenario, on the other hand, if you to scan every compartment for her before boarding, the two outcomes are hardly independent, so the chances of meeting revert to the 1/7 probability of her being on the train (or perhaps 1/7 x 1/2 if there is a 50% chance of the driver leaving you running around the platform).

Another popular myth that is easily explained is that some mystical force causes supposedly random events to occur in quick succession after a lengthy lull.

The fact is, there is nothing else they *can* do. There is no concerted guidance of such occurrences; neither from above nor from any lower level conspiracy. On the contrary, it actually is the *randomness* that causes them to bunch up.

Let us take a line 10 cm long and on it mark off 10 points at random, in other words - without order, regularity or pattern of any kind. The last thing to be regarded as patternless would be equal (1 cm) gaps. So, among the unequal gaps some are bound to be greater than the average of 1 cm; some only a little greater, one or two even much greater. When we come to insert the remaining marks we find that we have to 'squeeze' them in, in close successions, because they have to go *some*where. (The only other possibility, namely 'long, short, long, short, etc.' is a *pattern* too, i.e. not a random).

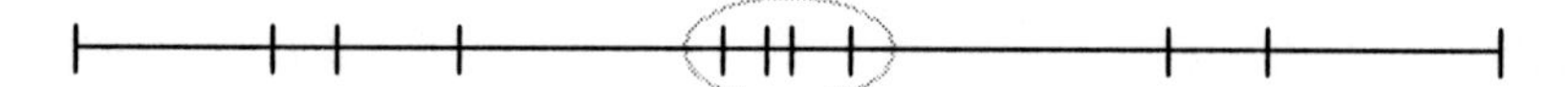

All this means is that when a given number of outcomes have to occur within a given period, occurrences in quick succession following a lull does not imply that anything in particular caused this to happen. It implies that the "anything" does not care a bit.

5.17 A serious warning about a health warning

Let us now turn to the crucial importance, when evaluating probabilities, of correct incorporation of all existing knowledge. (It is quite alarming how few of those who should know the following actually do.)

Let us imagine that you undergo a test for a fairly rare disease which on average only one in a thousand people contract. The test has a record of 99% reliability, i.e. an error occurs in only 1% of tests. The bad new is that your test report is positive (i.e. 'disease found').

What does this mean for you? How likely is it that you actually carry the disease? If you agonise, believing the probability to be 99%, your distress is more likely to be caused by ignorance than by the disease. The probability that you have the disease is actually 9% - 1 chance in 11! Ten out of eleven people diagnosed as positive are perfectly healthy. Why? Well, probably because they were careful, but why this result with a 99% accurate test?

Interpreting the test result as a 99% probability of carrying the disease results from not using all available information. In this case the test reliability information was regarded (partially), but the information about the *incidence of the disease* (one person in a thousand) was ignored and played no role in the appraisal.

Here, then is how the correct probability should be worked out, using *both* data. First, though, let us clarify exactly which probability we seek:

"What is the probability that receiving a positive result means that you are indeed ill"

The *happenings* are: "receiving a positive result", and the *specific outcome*, the probability of which we seek, is: "actually are ill (among those who received a positive result)".

The required probability is then the ratio between the number of ways of getting this *specific outcome*, and the number of ways to get the above defined *happenings. All* the ways!

Let us say that 100,000 people are tested.

Consider the number of the *specific outcomes*: of the 100,000 tested, 1 in 1000, namely, a 100, have the disease, and 99 of that hundred were correctly diagnosed by the test. So the total number of people who were tested positive *and* have the disease is 99.

Now we must establish how many people were tested positive in all (the *happenings*). We were aware, of course, of the above unfortunate 99, *but have overlooked a lot of others*: Of the 99,900 healthy people who, too, where subjected to a not entirely accurate test, 99% *were* correctly tested negative, which is very nice, but that still leaves 1% of healthy people who were wrongly tested positive. Only 1%, but 1% of 99,900, which is 999. The total number of 'positive' results is thus 99 **+ 999!**

And so, the real probability of you having the disease, even though your test was positive, is

99 / (99 + 999) = 99 / 1098 = 0.09 = 9%, or approximately one in eleven.

All this is to say that if a disease is rare, most of the positive test results will be of the 1% of wrongly diagnosed *healthy* subjects, because there are so many of them, and not of the few correctly diagnosed diseased ones.

Next time you go to your doctor for a check-up try checking *him* up on this one.

5.18 How good are the goods?

Now for the dessert we promised: how to evaluate the 'spread' of numbers around their average.

First we must consider the average itself: what it is, and how it is calculated. For a collection of various numbers, the 'average' is a single number which, by being "somewhere near the middle", can fairly represent *all* the numbers.

Let us say the numbers are prices of various items, but that for the sake of convenience we do not want to charge different amounts for different items. Instead, we want to establish the amount of a *single* price that is representative for *all* the products; we do not want to be out of pocket or overcharge, so the total that is received by charging this amount equally for the various products must equal the sum of the originally different prices of these products.

Let us say there are 9 items. We can find this single, representative price - the *average* price - by adding up all the original prices and dividing the total by nine. We know this provides the *average* as defined above, because if we multiply it back by 9 (which is the same as charging this average amount for each of the 9 items) we get back the right total of the original prices.

In another example, *n* children each owns a number of elephants; some have more, some have fewer: most unfair. The *average* in this case is the number of elephants each child would have if they all had the same number (the total number of elephants remaining unchanged). The easiest way to equalize this elephant ownership is to collect them all (or just add up their numbers; some *do* find arithmetic easier than moving elephants), and then divide the sum by *n*, the number of children. (If we, or rather the elephants, are lucky the result is a whole number.)

Thus the average of the seven numbers 3, 7, 8, 6, 6, 4, 1 is their sum (35) divided by 7 (=5). If the numbers are $A_1, A_2, A_3, \ldots, A_i, \ldots, A_n$ ('A_n' - the nth number, is the last one) we can use our now familiar shorthand for writing their sum as

$$\sum_{i=1}^{i=n} A_i$$

and the average is then

$$\sum_{i=1}^{i=n} A_i \; / n$$

The series 4, 6, 5, 5, 6, 4, 5, too, has 5 as an average, and so does 1, 2, 1, 16. In the latter case the average, 5, still correctly represents the 'combined effect', e.g., paying the price 5 four times amounts to the same as paying 1 + 2 + 1 + 16, but in any *other* sense this average, 5, does not really *look* very representative of the constituents, (in the way that barbed wire does not make a good representation of either snakes or toothpicks: it is somewhere in between, but not really much like either).

As the average of widely spread elements does not really amount to a faithful representation of the individual elements, it would be useful to have some *measure* of how well the average actually represents those elements. The simplest way in which to present the spread would be to add up the 'deviations' of the elements from the average, i.e. the difference between each element and the average, (and the smaller this sum, the more 'faithful' of the average). Two things must be noted, though:

1. The 'direction' of the deviation must be ignored. For instance, if the average is 5, both 7 and 3 must be considered to be 2 away from the 5, rather than +2 in the case of the 7 and - 2 in the case of the 3. If this is not observed the sum of all the deviations will always be zero (try it on the examples given above). It follows from this, incidentally, that another definition of '*average*' is "the number from which the sum of the deviations in both directions is zero".

(Disregarding the sign of a number e.g. using -2 as if it were +2, is called using its *absolute value*. The notation used to signify this is enclosure by square brackets: [a] means that the value of +a is used irrespective of whether 'a' is positive or negative).

2. Naturally, the larger the number of elements, the larger the sum of deviations; this is misleading, because we want the sum to reflect the breadth of the spread irrespective of the number of elements. To compensate for the effect of the number of elements on the sum of deviations, we can simply divide the sum by the number of elements. Taking 4 and 6 (average 5), the sum of the (absolute value of the) deviations is 2 which we divide by the number of elements (2), giving 1. If we have 4, 4, 6, 6, the average is still 5, the sum of deviations is 4, which divided by the number of elements (4) is again 1. This was to be expected, because in both cases the spread was between the numbers 4 and 6.

When we added some numbers and divided the sum by how many numbers there were, we called the result "average number". When we add the *deviations* and divide their sum by how many deviations there are, we call the result "average deviation".

The average deviation in the above example of 3, 7, 8, 6, 6, 4, 1 (with the average 5) is:

$$[3-5] + [7-5] + [8-5] + [6-5] + [6-5] + [4-5] + [1-5] = 14.$$

When we divide this sum by the number of elements (7) we obtain the average deviation: 2

In the other example 4, 6, 5, 5, 6, 4, 5 (with the same average) the deviations are

1, 1, 0,0, 1, 1, 0, their sum is 4, and the average deviation is 4/7 =0.571, which, as expected, shows a much closer 'fit' to the average.

In the example of 1, 2, 1, 16 the sum of the deviations is 22, the average deviation is 22/4 = 5.5, which is greater than the average itself (5). When an *average deviation* exceeds (or even approaches) the average it is useless.

Grown-ups actually use a more elaborate way of evaluating the spreads. Instead of summing the deviations, they first multiply each deviation by itself, then sum these selfmulted ("squared") deviations; next the sum is divided by the number of elements, and finally the *2-way fragment* (ex- "root") of the result is taken; (the self-multiplication of the deviations made everything rather big; taking the *2-way fragment* at the end makes the result small again). When all this has been done the outcome is called *standard deviation.*

In the above case of 3, 7, 8, 6, 6, 4, 1 this comes to 2.27 (check this by following the procedure described above), as opposed to the *average deviation* of 2.

There are three reasons for selfmulting ("squaring") the deviations. Firstly this takes care of the negative deviations: when a negative number is multiplied by another negative number the result is positive (for reasons that are explained in the next chapter). Secondly, it rightly penalises gross excesses: when one deviation is 10 and another is 2 their ratio is **5**:1, but the ratio between their *selfmults* (squares) 100 & 4 is **25**:1. Thirdly, by using this *standard deviation* some very specific useful conclusions can be drawn about the data, as will be described the in the following example. However, the reason why this *standard deviation*, namely, the way it is computed, leads to the following conclusions, is outside the scope of this text. So, the example of this application will have to be given without explaining why it works the way it does.

Let us say that you make shoes for the Chinese adult market. You want to minimise the number of sizes you manufacture while still catering for most of the Chinese population. You need to know what their foot sizes are, but cannot go out and measure them all. Therefore you take a 'sample': measure the foot size of a number of Chinese chosen at random. Having worked out the average size, let us say 21 cm, you then compute the standard deviation, as above, (except that you divide not by the number of measurements, but by that number less 1. The reason for this, again, cannot be explained here). Supposing the *standard deviation* you find is 1.5; you can then be reasonably confident that about 68% of the rest of the Chinese population will have foot sizes in a range of: {the average *minus* the standard deviation} and {the average *plus* the standard deviation}, that is: between (21 -1.5) and (21 +1.5) i.e. between 19.5 and 22.5cm. The range of the average +/- *twice* the standard deviation, i.e. 18 to 24, would cover 95% of the population.

Most of us do have some idea of what the word 'average' is about - an intolerable situation for those bent on confusing the healthy minds of young and old; so this time they devised something *really* crafty: instead of just concocting an unfamiliar, convoluted string of syllables, they took a word that already *had* at least *three different meanings*; and, were this not confusing enough, one of whose manifold of meanings is the word used for the concept of 'meaning' itself...

The *means* they chose to obfuscate what 'average' *means* was to use the word '*mean*'. Very *mean*.

Chapter 6

A Few Why S?

6.1 Misled by being shown how without why

The only things many of us (including you?) remember of our school maths are long multiplication and division, perhaps also that "minus times minus is plus", and perhaps... that it is best to leave it at that...

Sadly, even those who can still use these few residues of their maths education have no idea why they are done in the way they are. Why, for instance, we start long multiplication from the right, and long division from the left; why $-3 \times -2 = +6$. "That is how it is done..." probably sums up all the explanation we were given.

Most people no longer need to *do* these things: it is cheaper to buy you a calculator than the sustenance you need while doing the sums yourself. However, understanding the *reasons* behind the methods is still important. The main purpose of going into these basic topics here is to demonstrate how easy it is to explain and understand those 'untold reasons', and liberate one from the submissive blind obedience in learning maths.

Then we have circles and a certain letter π, and some procedures for dealing with fractions, all, apparently heaven sent. They are nothing of the sort. Heaven was too busy designing brains for people to use in pondering why heaven bothered.

Let us start with some basic arithmetic.

Multiplication and division from left and right

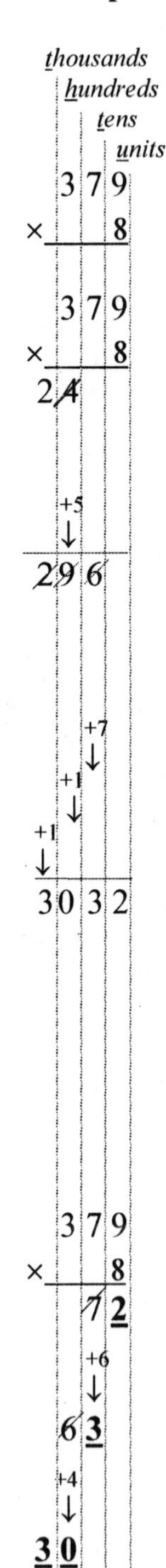

Why do we start multiplying 8 with the 9 - the rightmost digit in 379, and not with the 3 on the left? The way to look into this question is to see what would happen if we did it the other way round:

8×3 *hundreds* = 24 *hundreds*, i.e. 2 *thousands* and 4 *hundreds*, so we put 2 in the *thousands* column and 4 in the *hundreds* column.

Now we continue with 8×7 *tens* = 56 *tens* = 5 *hundreds* and 6 *tens*. We therefore insert 6 in the *tens* column, but then, when we come to put 5 in the *hundreds* column we find that the space is already occupied by 4; we therefore need to change it to 9 by adding the newly arrived 5.

Finally we do the 8×9 = 72, so 2 goes to the *units* col. and 7 to the *tens* column, but this column already has 6 in it, so it has to be changed: adding 7 to this 6 gives us 13 (*tens*) - 3 *tens* and 1 *hundred*; so 3 goes to the tens column and 1 to the *hundreds* column *which changes this column **again***: the new 1 added to the 9 already there = 10 *hundreds*, of which the 0 goes to the *hundreds* column and the 1 to the *thousands* column, which means that *another* column that was already occupied must be changed, by adding 1 to make it 3.

This shows that we do obtain the correct result also if we start from the left, but doing it this way we keep moving back and forth and must *repeatedly* alter the columns' contents.

Now let us see what happens when we start from the right:

8×9 = 72. 2 goes into the *units* column *which will not be touched again*, and the 7 is held in the *tens* column 'till further notice'. The next stage is 8x7 =56 *tens*, which together with the already present 7 *tens* is 63 *tens*: 3 *tens* are entered in the *tens* column *where they remain unchanged*, and 6 *hundreds* are held in the *hundreds* column. Lastly we have 8×3 *hundreds* = 24 *hundreds*, which together with the 6 *hundreds* already present is 30 *hundreds*: the 0 goes to the *hundreds* column - for *good*, and the 3 to the *thousands* column, for *best*.

Thus the advantage of starting on the right is that first the *units* are dealt with - permanently. Then the *tens* are settled with a single amendment, next the *hundreds* etc. The progression is in *one direction only*. The first and last positions are accessed only once and the others are amended only once. (Compare the number of amendments in the above two ways).

Now let us see how the same to and fro movement is incurred if long division is done the wrong way round, and is avoided when done the right way, which in this case is from left to right. Let us start with 785 divided by 3:

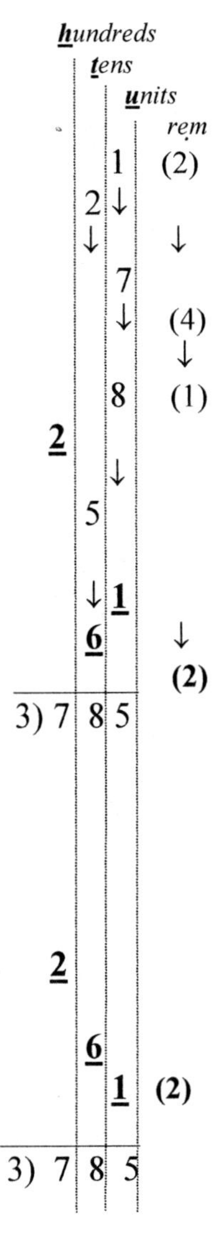

If we start from the right we have 5/3 = 1, remainder 2 *(units)*;

Next the 8 (*tens*) are divided by the 3, giving 2 *(tens)* which go to the *tens* column, and remainder 2*(tens)* which are divided by 3 to give 6 *(units)* which are added to the *units* column (which already contains 1 and is therefore increased to 7), and a remainder of 2*(units)* which, together with the already existing *units*-remainder, make 4*(units)* which, divided by 3, gives another 1 to the *units* column, raising it from 7 to 8, and a new *unit*-remainder of 1.

Last, we divide the 7(*hundred*) by 3, giving 2*(hundred)* for the *hundreds* col. and a remainder of 1*(hundred)* which, divided by 3 gives 3 *(tens)* which are added to the 2 *(tens)* which already occupy the *tens* column, storing a 5 *(tens)* there, and a remainder of 1 *(ten)* which is divided by 3 to give 3 *(units)* that need to be added to the 8 already in the *units* col., giving 11*(units)* which means 1 in the *units* col. (3rd change!) and 1 added to the *tens* col., changing it, from 5 to 6, and there is another *unit*-remainder of 1, giving a total *unit*-remainder of 2.

The result (261, rem.2) *is* correct, but what a mess getting there (like driving the wrong way down a one way street…).

Were we to start from the left, we would divide the 7 *(hundred)* by 3, giving 2 *(hundred)* + remainder 1*(hundred)*; the 2 goes in the *hundreds* col. and remains there unchanged. Then the 8 *(tens)* together with the remainder 1 *(hundred)* makes 18 tens, which, divided by 3 gives 6 *(tens)* which are entered in the *tens* col. - for good. (This step happened to leave no remainder.) Finally the 5 in the *units* col. is divided by 3, giving 1 + remainder 2, so 1 is entered in the *units* col. and that is that.

Does that not seem a good enough reason for doing it the right way? And is it not worth knowing?

6.2 Enemies of enemies

Why is $-1 \times -1 = 1$?

(Our purpose in bringing this up is not just to explain it, but to demonstrate how *anyone* could tackle this kind of task on one's own.)

One starts by trying to find a perceptible meaning for the *ingredients*. Let us begin with the 'minus'. To have "minus £50" means that adding £50 to your estate would bring its balance to 0. At the same time, the only way to get a 0 balance after adding £50 is to have had a *debt* of £50. So, "-£50" must mean "*debt* of £50". Now, a *debt* of 50 is what we regard as the opposite of *having* 50, that is, (+)50. Therefore, we may regard 'minus' as meaning the 'opposite'.

Then there is the ×, or '*times*'. We have already seen that when we say "give me 6 *of* those..." we mean "6 *times*...". So '-1 times -1' translates into 'the opposite (one) *of* the opposite'.

An *American Sunday buffet brunch* is absolutely the *opposite* of what the doctor *meant* when he recommended brunch instead of breakfast *and* lunch. What *he meant* was an ounce of bran and a leaf of decaffeinated lettuce, which is very much the *opposite* of an *American Sunday buffet brunch.* In short, the opposite of the opposite of something is the something itself. (A leaner example: *short* is the opposite of *tall,* and the opposite of *tall* is *short.*)

Let us take now the sentence "the opposite(one) of the opposite(one) of A is (one of) A"
(the 'ones' in the brackets
do not effect the contents) and
translate it into mathematical terms: $- \quad 1 \quad \times \qquad - \quad 1 \quad \times A = (+)1 \times A$

Bringing the trees a little closer to make out the forest:

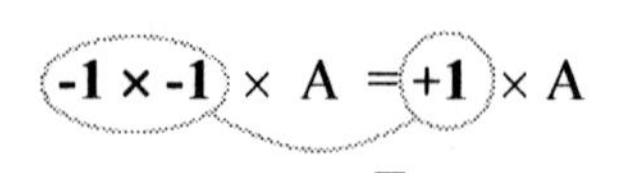

This is why. Not because you are vaguely sure that the teacher said so...

Two important corollaries can be drawn from the above result.

The above result can be generalised (somewhat loosely, but nevertheless correctly) as:

Two 'minuses' multiplied together 'eat each other up'. So, $-A \times -B = (+) A \times B$.

We know that "$P \times P = Q$" means that P is the (2-way) fragment of Q ('root' of Q). Now we found that also $-P \times -P$ equals (+)Q, so, also -P is the (2-way) fragment of Q ('root' of Q). This means that every (positive) number has a **pair** of *2-way fragments* ('square roots').

For example, the full answer to "what is the 2-way fragment ('sq. root') of 4" is not just 2, *but also -2.*

What then is the 2-way fragment ('sq. root') of -Q?

Neither $P \times P$ nor $-P \times -P$ yielded -Q, so -Q cannot have a 2-way fragment ('sq. root').

(Contrary to common belief, mathematicians have a lot of imagination, and so they pretend that negative numbers *do* have a 2-way fragment ('sq. root') and they call this an *imaginary number.*)

6.3 Bake your own Pi

If you go on a walk which always keeps you at exactly the same distance from some fixed point (which we call the 'centre'; 'o' in fig.1) you are walking in a *circle*. This fixed distance from the centre ('r' in fig.1) is the *radius* ('radiate': 'send out in all directions', in this case- the 'equal distance'). The distance across the circle is the *diameter*, or 'd' for short ('dia-': 'across'). In fig.1 this is line K-M, which leaves no doubt that d = 2r. *Circumference* is the shortest word they could devise for the length of the way around the circle; 'c' will do here.

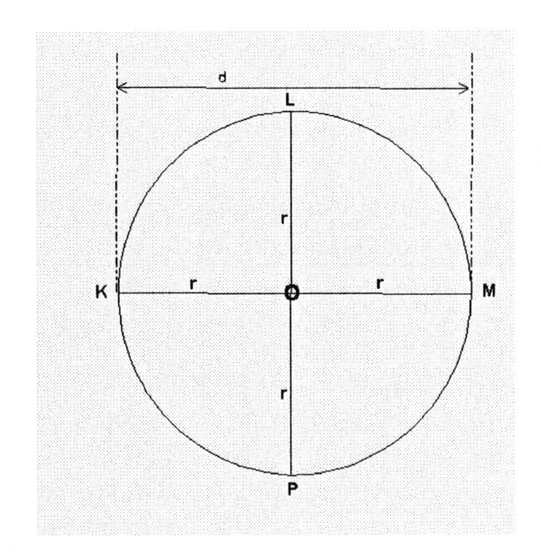

If you know d, how much approximately, is c?

With no one to ask, how would one go about finding out? Getting one to dare attempt anything at all is the object of this exercise, rather than just learn about circles.

Here is a useful way to estimate quantities for want of a more exact method.

Establish a known quantity which you are certain is *smaller* than the one you want to estimate, and another quantity that is certainly *larger* than it. The required quantity lies somewhere between these 'lower and upper bounds'. The closer to each other these are, the more accurate the estimated quantity will be.

Circumference c goes from K to M via L, and back to K via P. A definitely shorter route from K to M and back to K, would be going directly via o, i.e. twice the length of d. So, we know that the length of 'c' is larger than 2d. (2d is thus what we call the *'lower bound'*).

Now, the 'upper bound': a quarter of c, namely the section K-L, following the circle, is shorter than K-o-L which is 2r (=d). So, the whole of c, ***4*** such quarters, is *smaller* than ***4*** times 2r, or ***4*** times d. (4d is thus our *'upper bound'*).

Since c is more than 2d and less than 4d it must be somewhere near 3d. In the round world of circles establishing the exact value of this 'somewhere near 3' is not a simple task. The difficulty is to be expected: measuring the length of a straight line *is* straightforward; not so a curve: one might approximate a curved line to a series of short straight (i.e. measurable) sections; their sum would approximate the length of the curve; the accuracy of the approximation would improve the more, shorter, sections are used.

Instead of measuring and summing, it is possible to use mathematical methods to *calculate* the length of the sections, then investigate what happens to the *sum* of these lengths as the sections are shortened (and their number increased), until they become *infinitely* short and *infinitely* numerous, i.e. find the value of this sum of 'endlessly many nothings'. To explain how this is done is beyond the scope of this book (and examples beyond the scope of political correctness).

The result, namely, the exact value of this 'somewhere near 3' is also beyond the scope of book. In fact, of any book: It is 3.141592... etc., this etc. having no end. A shorthand symbol for this number is therefore *particularly* welcome: it is the letter π (pi), the abbreviation of the Greek word *periphery.*

Therefore $c = \pi d$ or $2\pi r$ (because $d = 2r$).

Now that we have a simple formula for working out the circumference, let us turn to the area of a circle (fig. 2).

We begin by applying the same principle: look for two shapes the area of which are easily calculated, one definitely larger than the circle, the other definitely smaller. If there is anything the area of which we know, it is a square. So we look for a square that is larger than the circle and one that is smaller. Both squares are chosen so that their areas can be expressed in terms of the same thing that is used to define the size of the circle: its radius.

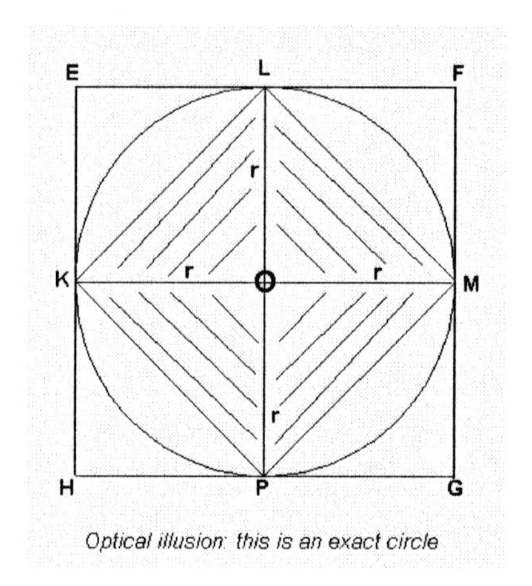

Optical illusion: this is an exact circle

The square that is obviously larger than the circle is EFGH. Its area is EF×EH, both of which can be related to the radius of the circle: EF and EH are each equal to 2r. Therefore the area of square EFGH is $2r \times 2r = 2 \times 2 \times r \times r = 4r^2$; and the area of the circle is less than this.

Now we must work out the area of the square that is smaller than the circle, square KLMP, again expressed by using r. Notice that the outer square EFGH contains 8 equal triangles such as ELK and KOL. The inner square KLMP contains 4 of these triangles, so its area is half that of the outer one, i.e. ½ of $4r^2 = 2r^2$. The area of the circle is therefore more than $\mathbf{2r^2}$ and less than $\mathbf{4r^2}$, once again “somewhere near **3**”. Might it be the same π as before (for the circumference)? Is there a connection between the area and the circumference? Indeed.

To see the connection, we regard the area of the circle as the sum of many thin concentric rings of (very small) width 'w'. If we cut open the outer ring and straighten it out we get a rectangle the length of which equals the circumference of the circle, $2\pi r$, and its width is w. Its area, and so also the area of the ring, is thus $2\pi r \times w$. The inner rings have progressively shorter length, the innermost being 0. On the average, the length of the rings is half-way between the outer and the inner, half way between $2\pi r$ and 0, namely ½ of $2\pi r$, that is, πr. The sum of the areas of all the rings is then their average length πr times the sum of their widths, which is simply r. So the area of the circle (any circle) is $\pi r \times r$, that is, $\boldsymbol{\pi r^2}$ (a little more than the $3r^2$ between the above squares).

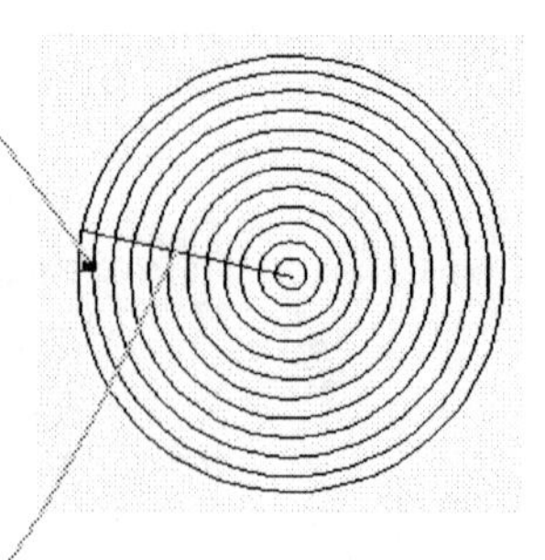

Note: the area of anything is always the multiplication of two lengths (or, as in this case, the multiplication of a length by itself). Similarly, the volume of anything is the multiplication of three lengths.

6.4 Common denominators uncommonly explained

Everyone remembers that one loses marks at school unless one uses 'common denominators' when adding fractions. Some even remember *how* to work them out. But how many know (or ever knew) the *reasons* behind the method used?

(The best proof of "knowing" is being able to explain it clearly to others.)

As a good start we better clarify what we are talking about, and so we will use *divider* instead of "*denominator*". We will also replace "*numerator*" by the "*divided*".

4 cats plus 3 cats is 7 cats, but 4 cats plus 3 mice, though it may produce some lively action, has no meaning in terms of a numerical answer. Our minds run into problems when we attempt to add (or subtract) unlike entities. In some cases, such as "5 miles + 7 acres", the addition is totally meaningless. In other cases there is a way out. For instance "50 inches + 6 metres" is not 56 of *anything*, but it still means something, because inches and metres are both units of length, and so is their sum. There remains the question of how we accomplish this sum.

As usual, the best way to tackle the problem is to identify the specific element that creates the trouble.

Let us run through the presented elements: we know what '50' and '6' are. We know what inches and metres are, and we have school certificates to testify that we know what '+' is. There is nothing, it seems, that we do not know, so where is the problem?

It is that inches and metres are not the same thing. This means that we would overcome the problem if we found something that could replace both 'inch' and 'metre', so that the numbers are followed by identical items.

This 'analysis' may seem childishly simplistic, yet it is surprising how many seemingly insurmountable problems can be easily overcome by using this simple kind of approach.

Let us consider first "50 inches + 6 foot": A simple way to get identical items following the numbers would be not to replace both 'inch' and 'foot' by something else, but to retain one of them, let us say 'inch', and try to replace 'foot' by a certain number of inches:

One foot is 12 inches, so the "6 foot" becomes 6× 12 inches (72 inches), and so the above "...inches +...foot" becomes the 'addable' 50 inches + 72 inches = 122 inches.

With the first example, 50 inches + 6 metres, this cannot be done because unfortunately (*very* unfortunately) a metre is not equal to a simple number of inches; in fact the fraction, as in π, stretches endlessly. Therefore we do need a third unit for which we know the *exact* number of times it fits into one inch and into one metre: the great little millimetre. It goes 25.4 times in one inch, and of course 1000 times in one metre.

50 inches + 6 metres can therefore be rendered as:

50× **25.4mm** + 6× **1000mm**, or 1270 mm + 6000 mm.

Everything is in mm, including the result: 50 inches + 6 meter = 7270 mm.

How do we apply this principle to $\frac{4}{5}+\frac{2}{3}$? How much is this?

We could, of course, convert these fractions to decimals ('that thing with the point') by long division:

4/5 = 0.8; 2/3 = 0.666... and then add the two decimal numbers to give 1.4666...

There are good reasons, however, for not doing this. One is that the decimal equivalent of a fraction is often an infinitely long string of digits, as in the above 2/3. If we convert a fraction, or the sum, into one of these endlessly long decimals, we must decide where we 'cut off the tail', in other words lay down the degree of accuracy of the number. Often however the required accuracy can only be known later, depending on the application. Leaving the figures in the form of fractions, and doing the sums as fractions, they retain full accuracy.

There are also situations in which it is easier to "see what is going on" in simple fraction form: it is hardly obvious at a glance, for instance, that 21×0.142857 is 3. This is much more apparent when 0.142857 is left in its fraction form of 1/7 ($21 \times 1/7$, or 21/7).

How then do we add 4/5 and 2/3?

Firstly: $\frac{4}{5}+\frac{2}{3}$ is *not* equal to $\frac{4+2}{5+3}$, even if *"it looks like it"...*

This kind of (non) reasoning rarely leads to correct results, particularly in maths, and even when, by luck, they are correct, we still do not know what we are doing. We *do* want to know what we are doing.

Our purpose here is to trace the route we would, or ought to, take if left to our own devices. Clues as to what to do can only come from considering the ingredients themselves, in this case -

$\frac{4}{5}+\frac{2}{3}$. Now, $\frac{4}{5}$ is the same as $4\times\frac{1}{5}$, namely '4 fifths*'. Similarly $\frac{2}{3}$ is '2 thirds'.
"4 fifths and 2 thirds" reminds us of the problem - *and the way out of it* – that we had with "50 inches and 6 metres", namely of adding dissimilar items.

Let us then try the same solution: replace 'fifth' and 'third' with a single, common, thing. 'Fifth' and 'third' come from the dividers 5 and 3; we must therefore find a single substitute for these two dividers.

Let us suppose we use 10, a nice number; this would give us '4 *tenths*' + '2 *tenths*', which are easily added to give 6 *tenths*. Easy it is, but the problem is that we are not adding what we were given: '4 tenths' is not '4/5', and '2 tenths' is not '2/3'.

Obviously, if we change the divider (which, we decided, we must) a change must also be made elsewhere in order to bring the values back to 4/5 and 2/3. As to 'elsewhere', we

---

* Just as, remember, $4\times K$ is "4 *of* K" or- four K's , and 1/5 is called a fifth.

have little choice: it must be the 'divided' numbers (- above the dividing line). By analogy - if you want to divide a cake between a greater number of children without reducing the size of the slices the only solution is to increase the size of the cake. Since 4 is now divided by something bigger, we must also increase the divided 4. But how?

Intuitively, by doing the 'same thing' to both the divided and the divider. But *what* 'same' thing? Changing the divider from 5 to 10 could be done by *addition* (of 5) or by *multiplying* (by 2). So, do we also *add* 5 to the divided 4, changing it to *9*, or do we *multiply* the 4 by 2, changing it to *8*? The first case would change the 4/5 fraction to *9*/10 while the second to *8*/10. Try to find out, using long division if necessary, which of these possible results is in fact equal to 4/5.

You will find that compensation by equal *multiplication* is the one that works. We have encountered this before: the value of a fraction is preserved when both the divider and the divided are multiplied by the same number. (Like the great 'sales': double prices on the day before, then display the big 'half price' posters...)

Thus we introduce our common divider in the 4/5 by replacing it with 8/10. So far so good. Now we must perform the same operation on 2/3, i.e. replace also the 3 by 10. This must go together with multiplying the divider 2 by the same multiplier that turns 3 into 10. At this point, when everything seemed to be going so smoothly, our system collapses, for the multiplier of 3 that gives 10 is 3.33333....

Avoiding 'unlimited' disasters of this kind is the whole purpose of this exercise. 10 was nice, but not nice enough, because what we need is a number that can be reached from *all* the dividers, in this case from the 5 *as well as the 3*, by multiplying them with exact, preferably *whole* numbers. How?

Here comes a really nice trick:

Let us call the two dividers G and H. We want a number that can be reached from *both* G *and* H - in each case through a multiplication by some whole number. In other words, we want to find a whole number to multiply the G with, and another whole number to multiply the H with, so that in both cases we get the same result. And we want a general rule - one that will work with *any* pair of dividers. (This is the reason for using G and H instead of 5 and 3.)

It is at this stage that a teacher ruins the lesson if he does anything other than let the class grope for the solution by themselves, because this is what learning maths is all about: not "doing sums", but figuring out *how* to do them. So here is *your* turn (turn to think before turn the page.)

The 'trick' is simply this: multiply divider G by divider H, and divider H by divider G!

(The resulting G × H will be found to be very equal to H × G!)

With this principle established, let us return to our $\frac{4}{5} + \frac{2}{3}$:

5 (the divider in 4/5) is multiplied by 3, and the same of course we do to the (divided) 4:

$$\frac{4}{5} = \frac{4 \times 3}{5 \times 3} = \frac{12}{15},$$

3 (the divider in 2/3) is multiplied by 5, and the same of course we do to the (divided) 2:

$$\frac{2}{3} = \frac{2 \times 5}{3 \times 5} = \frac{10}{15};$$

-in this way we get both the divided numbers sitting on 15, which we call the "common divider".

and so, the sum $\frac{4}{5} + \frac{2}{3}$ can thus be written as $\frac{12}{15} + \frac{10}{15}$ which *can* be added straightforwardly*,.

because it spells $12\,\textit{fifteenths} + 10\,\textit{fifteenths} = 22\,\textit{fifteenths}$, or $\frac{22}{15}$.

$$\text{So, } \frac{4}{5} + \frac{2}{3} = \frac{22}{15}.$$

By all means succumb to the temptation to check this (by using long division to convert all the fractions into decimals).

6.5 Crushing ladders

We know what $\frac{A}{B}$ means, but what about $\frac{\frac{A}{B}}{C}$, $\frac{A}{\frac{B}{C}}$, $\frac{\frac{A}{B}}{\frac{C}{D}}$?

These clearly appear to be examples of multiple divisions.

(Note that in contrast to the previous 'common divider' section, here the elements of division (*A, B, C, D*) can be decimal numbers.)

* what a long word to express *'straightforwardness'...*

An important point must be made at the outset here.

In a succession of additions or multiplications such as $3 + 5 + 6$ or $4 \times 6 \times 3$, we get the same result irrespective of the order in which the numbers are added or multiplied. We arrive at the same 14 whether we first add 3 and 5 (= 8) and then add 6, or if we first do 5 + 6 (= 11) and then add 3. Similarly, 4×6 (= 24) multiplied by 3 results in 72 as does 6×3 (= 18) multiplied by 4.

In multiple subtractions or divisions however, the result varies (as during courtship) according to the order of the steps.

Let us take 10 - 7 - 1:

> if we do 10 - 7 (=3) first, and then subtract the 1, we are left with 2;
>
> if we do the 7 -1 first, and *then* subtract the result (6) from the 10, we get 4.

The order of operations therefore does matter. The same applies when we divide 24/6/2:

> if we start with 24/6 (= 4) and then divide by 2, we end up with 2, but doing the 6/2 (= 3) first and *then* dividing the 24 by this 3, produces a much larger result: 8.

Thus there is little point in setting out a series of subtractions and divisions without specifying the order of operations. One common way in which this is done is to put the part to be done first in brackets, e.g. 24/(6/2), which means doing 6/2 (=3) first, and then dividing the 24 by it, (giving 8).

In the absence of brackets, the convention is to do things in the order in which they are written, but when divisions are set out in the vertical 'ladder' form, the order is indicated by the length of the dividing lines: items separated by the shorter lines are divided first, and their results are then used for the divisions of the longer line. For instance:

$\dfrac{\frac{24}{6}}{2}$ means (24/6) /2 i.e. 4/2 (=2) , while $\dfrac{24}{\frac{6}{2}}$ means 24/ (6/2) i.e. 24/3 (=8).

If the ladders contain numbers the rungs can be removed simply by carrying out the divisions as above, but how is this to be done if the rungs contain symbols (standing for any numbers), as in

$$\dfrac{\frac{a}{b}}{c} \quad \text{or} \quad \dfrac{a}{\frac{b}{c}}$$

The first question to ask is why should we *want* to remove rungs of the ladder?

Writing these sums as simple 'one rung' fractions makes it easier to see what is going on, that is, what contribution the various elements make. For instance, if the C in the examples above is increased, does that increase or decrease the value of the whole thing?

Having decided that it is desirable to use only one rung, we must decide how to make the change, and more importantly, *what method we use to find a way of doing so.* As usual we will begin with something similar the meaning of which we *know*, identify what is different in the case at hand, and establish the effect of this difference.

We will assume that B, C, D represent numbers that are not smaller than 1. (If they are smaller we would end with the same conclusions, but the stories are longer.)

Let us start with $\frac{\frac{A}{B}}{C}$.

The length of the lines here tell us that we do $\frac{A}{B}$ first and then divide the result by C. The question now is how the further division by C affects the result of dividing A only by B?

The answer is that it further reduces the result, making it yet C times smaller. Now we must ask ourselves by what other means, one that uses only *a single* rung, we might achieve the same effect of making the result C-times *smaller* than that of A divided only by B.

To diminish the result of a division, one increases the divider. To make the result C*-times* smaller the divider (the B, in this case) needs to be made C-times larger, i.e. A gets divided by $B \times C$.

So, $\frac{\frac{A}{B}}{C}$ is the same as $\frac{A}{B \times C}$

Now let us turn to $\frac{A}{\frac{B}{C}}$.

Here the lengths of the lines tell us that the first thing that happens is division of B by C, after which A is divided by the result of B/C. What is the effect of all this compared with the simple division A/B?

A is now divided not by B but by something smaller: C times smaller. When a cake is divided between fewer children, the portions are larger. Specifically, if there are C-times fewer children the portions get C-times larger. So, when A is divided by something that is C times smaller than B, the result of this division is C times larger than it would be were A divided by B alone.

As before, we now look for a way to do the same thing using one rung only, namely, writing a one-rung fraction that is C times larger than $\frac{A}{B}$.

One way to describe how we removed a rung in the previous case is the replacement of one of the divisions by a multiplication, (- multiplications do not use rungs). Using multiplication ties in with the observation that there is another way to get a (C-fold) increase of the (cake's) portion: instead of reducing the number of children (dividing their number by C) we can make the (divided) *cake* larger, C times larger. So, similarly, instead of dividing the A by (the 'reduced') B/C, we divide (an 'increased') $A\times C$ by B, i.e.:

$$\frac{A}{\frac{B}{C}} = \frac{A\times C}{B}.$$

Note: if $A = 1$ we obtain the useful result $\dfrac{1}{\dfrac{B}{C}} = \dfrac{C}{B}$ e.g. $\dfrac{1}{\frac{1}{2}} = \dfrac{2}{1}$ (=2) i.e. $\dfrac{1}{half} = two.$

We now have the opportunity to deduce a very important result:

If C is very large, B/C is very small. So, in the ladder $\frac{A}{\frac{B}{C}}$ the A is divided by something very *small.*

At the same time, the very large C makes the (equivalent) $\frac{A\times C}{B}$ very *large*.

To summarise: if A is divided by something very small the result is something very large.

If we read "very" as 'very very *very*', and take it to its extreme, 'very very *very* small' becomes 0, and v. v. v. large become 'infinity', a term for "endlessly large", represented by the symbol ∞.

Therefore, $\dfrac{anything}{0} = \infty$.

As no one got to infinity and came back, this result is best avoided, and so is the division by 0 !

Finally, $\dfrac{\frac{A}{B}}{\frac{C}{D}}$:

The length of the lines demand that A/B and C/D be carried out first, and then that the result of A/B be divided by the result of C/D.

Again, the same line of attack: first we pretend there is no D, because we already know how to handle the remaining $\dfrac{\frac{A}{B}}{C}$ (i.e. write it with one rung only, as $\dfrac{A}{B \times C}$).

Next we examine the effect of letting D back into its place, dividing the C, i.e. in the

$\dfrac{A}{B \times C}$ we replace the C by C/D;

The C has a dividing role, and as it is now made D-times smaller, the effect, as before, is to make the whole thing larger - D times larger. We already know of another, a 'rung-free', way to make $\dfrac{A}{B \times C}$ larger: the required D-fold increase we achieve by *multiplying* the A by D, namely: $\dfrac{A \times D}{B \times C}$.

Summarizing the steps: $\dfrac{\frac{A}{B}}{\frac{C}{D}} = \dfrac{A}{B \times \frac{C}{D}} = \dfrac{A \times D}{B \times C}$,

Leaving out the middleman: $\dfrac{\frac{A}{B}}{\frac{C}{D}} = \dfrac{A \times D}{B \times C}$

(Refrain from memorizing this *'rule'* of "who goes up and who goes down", always *work it out* using the above simple considerations.)

Chapter 7

Generalisations and Shortcuts

Having confused students with Greek and Latin, why not throw in some Arabic too? That is why the word *Algebra* was introduced, and why we discard it. The reason is that this word too can be replaced by something that hints at what it is about.

If we drop the tens digit of 15 what is left is one third of what we started with.

No other number behaves in this way, so there is no generalisation to be made about the phenomenon.

76^2 however ends in 76 (5,776); 76^3 ends in 76 (438,976); so do all *selfmulted* 76s, that is to say, *selfmult* 76 with any *num.mults* ends in 76, for instance 76^5= 2,535,525,376. Therefore in this case we can formulate a generalised rule: "76^n ends in 76", i.e. that this statement is true whatever n is.

76 and 25 are the only two-digit (positive, whole) numbers for which changing the *num. mults* does not affect the last two digits of their *selfmults*. But don't start a new religion over this revelation, it is less surprising than it appears to be: if it is true for the square of a number, it is bound to be so for any *num.mults*.

Another example: (100 + 3) multiplied by (100 - 3) equals $100^2 - 3^2$

1003 × 97 = 10,000 - 9 = 9991

This holds for numbers other than 100 and 3; in fact, as we will prove later, it holds for any other numbers. For instance:

$$(40 + 1) \times (40 - 1) \text{ is } 40^2 - 1^2$$

$$41 \times 39 = 1600 - 1 = 1599$$

Therefore this can also be generalised, as follows:

$$(k + h) \times (k - h) = k^2 - h^2$$

Or, in words: "The result of multiplying the sum of any two numbers by their difference is the difference of their squares".

'*Any number*' cannot be expressed by numerals (*0 - 9*); that is why letters, such as the k and h above, are used.

The generalisation made above can also serves as a useful *shortcut*, a substantial one for the above multiplication. The subject of this chapter is

- generalisations that are useful in deriving conclusions from given information
- methods of shortcuts in calculations

So, we will call the subject *"Gens & Shorts"*. For short.

An important point concerning this subject: there are no 'miracles' involved - all that is done using (ex)"algebra" could be done without it, only the processes would be much more tedious.

> *Note*: We will call here items that are added to, or subtracted from, each other 'components', because components are what things are composed of. The perpetrators of 'logarithm words' use the term '*term*' (!?) instead of 'component*s*'; and they use the ambiguous term '*factor*' for what we call '*multiplier*'.

The more concise an expression, the easier it is to use. We cannot shorten 7×6 or $W \times E$ by dropping the $\times$ because 76 and WE mean something else. However numerals next to a letter, for instance 56Y, could be shorthand for *either* of $56 \times Y$, $56 + Y$ or $56/Y$: the $\times$ is chosen to be shortcut, probably because multiplications appear more frequently than other operators. The same shorthand can also be applied between a letter or number and a bracket:

$$A(B + 7) \text{ means } A \times (B+7) \quad \text{and} \quad 22(3 + Y) \text{ means } 22 \times (3+Y)$$

7.1 Some fun in and out of brackets

We need to start by establishing some relationships without which not much can happen here.

Four cats plus seven cats is eleven cats: $11 \text{ cats} = 4 \text{ cats} + 7 \text{ cats}$

If we write '(4 + 7)' instead of '11', this becomes: $(4 + 7) \text{ cats} = 4 \text{ cats} + 7 \text{ cats}$

By 'four cats' we mean the same thing as

'four times a cat', i.e. $4 \times \text{cat}$, etc. so the line above becomes:

$$(4 + 7) \times \text{cats} = 4 \times \text{cat} + 7 \times \text{cat}$$

Maths does not discriminate between animals; the above is true for dogs, lizards, for anything, even for 'C' in place of 'cat'. It is equally true for numbers other than 4 and 7, for instance for numbers A and B. So if we rewrite the line above, substituting

C for cat, A for 4, and B for 7, we have $(A + B) \times \boldsymbol{C} = A \times \boldsymbol{C} + B \times \boldsymbol{C}$

Any multiplication can be turned around; e.g. $4 \times 3 = 3 \times 4$,

so we can also have $\boldsymbol{C} \times (A+B) = \boldsymbol{C} \times A + \boldsymbol{C} \times B$

This is where the fundamentally important "brackets disposal method" ('bracket opening') comes from.

This should not be confused with $(A \times B) \times C$, which is just $A \times B \times C$. If in doubt try this with numbers such as:

$$(3 \times 4) \times 5 = (12) \times 5 = \mathbf{60} = 3 \times 4 \times 5.$$

[whereas $(3 + 4) \times 5 = (7) \times 5 = \mathbf{35} = 3 \times 5 + 4 \times 5$].
↑

Often, we find it useful to do the reverse of 'bracket opening':

For example, $91 + 14A$ (14A is the same as $14 \times A$)

can be written as $7 \times 13 + 7 \times 2A$;

the '$7\times$' can now be 'taken outside a bracket', and the expression then becomes

$$7 \times (13 + 2A), \quad \text{or, just } 7(13 + 2A).$$

What about $(A+B) \times (C+D)$?

What stands in the way of evaluating this is the brackets. The object of this exercise is therefore, not surprisingly, to write this *without* brackets.

Once again we will rely on our *'beat-the-problem'* method: begin with something similar which we *know* how to deal with, then look for a way to cope with the differences. A similar form that *is* familiar is $K \times (C + D)$. The difference is that now, instead of a single K multiplying the bracket (C+D), there is another bracket (A+B). So we will first convert the problem into the familiar form by replacing (A+B) with a single symbol ☠;

thus instead of $(A+B) \times (C+D)$

we write: ☠ $\times (C+D)$

Do not worry about the skull. This is only temporary.

We know how to write this without the bracket: ☠ $\times C +$ ☠ $\times D$

And we are already done with the skull: we replace it

by putting back the bracket (A+B), giving

$$(A+B) \times \boldsymbol{C} + (A+B) \times \boldsymbol{D}$$

All that remains to do now is to discard these two brackets too:

$$A \times \boldsymbol{C} + B \times \boldsymbol{C} + A \times \boldsymbol{D} + B \times \boldsymbol{D}$$

So we find that:

$$(A+B) \times (\boldsymbol{C}+\boldsymbol{D}) = A \times \boldsymbol{C} + B \times \boldsymbol{C} + A \times \boldsymbol{D} + B \times \boldsymbol{D}$$

We may describe this as "each item in one bracket multiplies each item in the other bracket, and all these multiplications are added together".

So, this is not because the *teacher said* so, but because otherwise $3 \times (19+5)$ would not be 72, and this *would* be serious.

Let us give an example of how this *generalisation* can be used as a useful *shortcut.*

It may be used for a 'trick' that allows the quick mental multiplications of two-digit numbers in which the 'tens' digit is the same in both, and their unit digits add up to ten, such as 72×78, 45×45, 99×91 etc.. If you know the method you would know at a glance that these are equal to 5616, 2025, 9009 respectively. Why?

Let us consider the example 72×78. 72 can be written as 70+2, and 78 as 70+8. So, using brackets the 72×78 then becomes

$$(70+2) \times (70+8).$$

This, as we know now (using the above 'each by each') can be written as

$$70\times70 \ + 2\times70 + 70\times8 \ + 2\times8.$$

Both middle components contain a multiplier 70 - these originate from the requirement that the original numbers have the same 'tens'. As we know these two middle components can be written as $(2 + 8) \times 70$, i.e. ***10*** $\times$ 70

(This is why the units, in this case 2 and 8 had to add up to 10).

So, the above now becomes:

$$70\times70 \ + 10\times70 \ + \ 2\times8.$$

Lastly, the first two components, the $70\times70 + 10\times70$

can be reduced to $(70+10) \times70$

$= 80 \times 70$

and our 72×78 ends up as $80\times70 \quad + \ 2\times8.$

This, with your hands in your pockets, is 5616.

The rule

The front digits: the common 'tens' $\times$ one extra ten; *The two rear digits*: the multiplied units.

Check the other two examples:

63×67: on the left goes $6\times (6+1) = 6\times7 = 42$; on the right goes $3\times7 = 21$;

together: 4221.

99×91: on the left: $9\times (9+1) = 9\times10 = 90$; on the right the two(!) digits: $9\times1=$ **09**;

together: 9009.

(Now you can go out and show off, but only if you remember for which numbers this works, and only if you can explain *why* it works!)

This promiscuity of "each member of one bracket multiplying each member of the other bracket" holds for any number of elements in the brackets. For instance

$(A+B+C) \times (D+E+F+G)$ yields a sum of 12 (3×4) multiplying pairs, all the possible pairs consisting of one element from each bracket.

When there are more than two brackets, for instance:

$$\underbrace{(A + B) \times (C + D)}_{\text{☠}} \times (E + F),$$

we start once again by substituting a skull for everything other than one of the brackets; in this case it may be for the $(A + B) \times (C + D)$, so that the multiplication above is reduced to ☠ $\times (E+F)$.

We know that ☠$\times(E+F)$ is ☠$\times E$ + ☠$\times F$, and we also know what to restore in place of the skull, and how to write it without brackets: $A{\times}C + B{\times}C + A{\times}D + B{\times}D$. So, each of the E and F multiply now a bracket containing this expansion of what was temporarily represented by the skull:

$$(A{\times}C + B{\times}C + A{\times}D + B{\times}D) \times \boldsymbol{E} \quad + \quad (A{\times}C + B{\times}C + A{\times}D + B{\times}D) \times \boldsymbol{F}$$

We also know how to get rid of *these* brackets... So we end up with a sum of 8 triplets of letters

$$A{\times}C \times \boldsymbol{E} \ + \ B{\times}C \times \boldsymbol{E} \ + \text{ etc.... } + \ A{\times}C \times \boldsymbol{F} \ + \ B{\times}C \times \boldsymbol{F} \ + \text{etc....}$$

A healthy exercise would be to expand both the

$(A+B+C) \times (D+E+F)$ and the $(A+B) \times (C+D) \times (E+F)$ and then replace the letters with numbers, for instance:

$(1+2+3) \times (4+5+6)$ and $(1+2) \times (3+4) \times (5+6)$, and check whether the result of the expansions tally with what you began with, namely 6×15 and $3 \times 7 \times 11$.

The difficulty here is not the maths but the patience, and the only scary stuff is the skull.

We now move on to three special cases which are of interest.

1. If you make a square with sides L, the area of the square is L^2. If you increase the length of the sides by s, i.e. make a square of sides L+s, what will be the area of this new square?

" Looks like $L^2 + s^2$...?" - Maybe it does, but the answer is wrong, and the approach even more so!

There are two ways in which we can work out this area correctly.

The first way is to *draw* a square with sides L+s and see what its area consists of: we find that it is made up of a square of area L^2, a square of area s^2, and of two rectangles of area L×s.

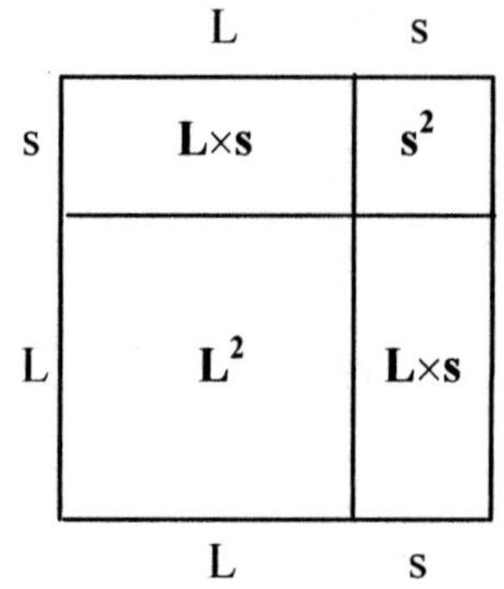

	L	s
s	**L×s**	**s^2**
L	**L^2**	**L×s**
	L	s

The second way is to *calculate* the area of this square with sides L+s.

If L^2 is the area of a square with sides L, then for sides '$\wp$' the area would be $\wp^2$, and for sides (L+s) the area is $(L+s)^2$. Let us work out what this consists of:

$(L + s)^2$ of course means $(L+s) \times (L+s)$

This, as we know, is $L\times L + L\times s + s\times L + s\times s$, which, in order to save ink, we can write

$$L^2 + 2L\times s + s^2$$

This is the same as the components of the drawing above: two square L^2 and s^2, and two rectangles L × s.

For those who prefer A and B to L and s this can be written:

$$(A + B)^2 = A^2 + 2A\times B + B^2$$

As before, this result can also serve to help with some mental arithmetic, but before showing how, one must stress the *importance* of the ability to make quick mental calculations and estimations. Calculators are invaluable, but those who use them exclusively lose their 'feel' for quantities, and become totally dependant on the erroneously assumed infallibility of their finger-work. If you are unable to look at a sum and know what sort of result to expect you better let someone else do your invoicing...and your weekly shopping.

Let us now see how the above result can help us with 104^2.

If we write 104 as the sum of two numbers, A and B, then 104^2 means evaluating $(A+B)^2$ which, as we know now, involves working out A^2, B^2 and (2×)A×B. To make it easy to calculate the squares as well as the A×B the A and B should be round figures or small figures, or, as in this case - one of each. Finding two such numbers, the sum of which is 104, is not *very* difficult...

Indeed, 100 and 4: the task of 104^2 is converted into $(100 + 4)^2$ and thus requires working out nothing harder than 100^2, 4^2 and (2×)100×4, all of which can easily be done (and added together) without a solar-powered calculator wasting the precious English sunlight.

$$104^2 = (100 + 4)^2 = 100^2 + 2\times100\times4 + 4^2 =$$

$$10{,}000 + 800 + 16 = 10816$$

Surely, you would not resist the urge to check that 203^2 say, equals 41209.

(200^2 =40,000; 2×200×3=... etc.)

2. For the next task we need to note that $6 \times (-5) = -30$:

This is because if we regard having £(-5) as *owing* £5, six times that unhappy state of affairs means *owing* £30, viz. *having* £(-30).

In general terms then, the rule is $A \times (-B) = -A \times B$ (or $-B \times A^*$).

Also: $-A \times -B$ can be written as

$-1 \times A \times -1 \times B$ (since '-A' is the same as "-one of A")

Changing the order: $-1 \times -1 \times A \times B$

$= (+)\ A \times B$ (we know all about $-1 \times -1 = +1$)

We are now ready to find what $(A - B)^2$ consists of.

The above means of course $(A - B) \times (A - B)$ which, stripped of brackets, and minding the minus symbols, is

$$A \times A + A \times -B + -B \times A + -B \times -B$$

or, in a shorter form

$$A^2 + 2 \times A \times -B + B^2$$

the $2 \times A \times -B^*$ can be rearranged to $-2A \times B$.

Thus $(A - B)^2$ is: $A^2 - 2A \times B + B^2$

Note that here too the B^2 has a + sign; the difference between the expansion of $(A + B)^2$ and $(A - B)^2$ is only in sign of the $2A \times B$.

3. Finally: $(A + B) \times (A - B) = A \times A + A \times -B + B \times A + -B \times B$

$$= A^2 - A \times B + B \times A - B^2$$

The two middle components conveniently undergo Mutual Assured Destruction, leaving us with the neat result:

$$(A+B) \times (A-B) = A^2 - B^2$$

Again this helps with calculations such as 52×48.

Any two numbers, N and M, say, can be written such that N is a *sum* of a certain pair of numbers, A and B, and M is the *difference* of the same A and B. However, the method is helpful only if these A and B are numbers that can be easily squared. In this case, where N is 52 and M is 48, this is so because A=50 and B=2 , namely:

52×48 can be written as $(50+2) \times (50-2)$. This, as we have just seen, is equal to $50^2 - 2^2$.

For the same money, would you not rather do 2500 - 4 than 52×48?

---

* Multipliers can always be rearranged, and '-W' is the same as $-1 \times W$.

We turn now to a remarkable phenomenon. A typical school survivor, told to

"do $(A+B) \times (A-B)$" will dutifully produce $A^2 - B^2$

(we are talking about the *good* ones here). Chat with him a while and then tell him to:

"do $G^2 - H^2$", and the proficient, tamed response will be $(G+H) \times (G-H)$.

Mice run back through a maze with the same alacrity they ran forward only if they are rewarded with cheese at both ends. But then mice, having been spared normal schooling, only do something when they know its *purpose*...

When, may *we* ask, is there a purpose in transforming $(S+Z) \times (S-Z)$ to $S^2 - Z^2$, and when is there a purpose in doing the reverse?

For example, given $(W+S) \times (W-S) + S^2$, changing the

$(W+S) \times (W-S)$ to $W^2 - S^2$ is nice, because together with the $+S^2$ all that is left is W^2.

On the other hand, given $(P^2 - Z^2) \ / \ (P - Z)$,

replacing $P^2 - Z^2$ with $(P+Z) \times (P-Z)$ gives $\frac{(P+Z) \times \cancel{(P-Z)}}{\cancel{(P-Z)}}$

which dissolves into $P + Z$.

7.2 Finding your bearings after a spin

Let us look at another example of a mathematical shortcut.

Consider a sentence such as "The driver did not deny that he refrained from avoiding the prevention of failure to miss flattening the skunk." Is the skunk dead or alive?

Fortunately most writing is not *quite* that bad, but often nearly so.

Since we know why $-1 \times -1 = (+)1$ we can make good use of it here, but first we must consider some *GEN*eralisations. We know that $-1 \times -1 = 1$, but what is $-1 \times -1 \times -1$?

The first '-1×-1' pair can be replaced by $+1$, so the $\mathbf{-1\times -1\times -1}$ becomes 1×-1, which is just **-1.**

Four (-1)s is equivalent to two pairs; each pair gives +1; the two pairs amount to 1×1, which is 1 (and was hardly worth the effort). So, $\mathbf{-1 \times -1 \times -1 \times -1 = (+)1}$.

Continuing with series of ever more (-1)s, we find that *selfmulting* an even number of (-1)s always yields +1, and that an odd *Num.mults* always yields -1.

Our purpose is *SHORTS*, so we can simply write

$(-1)^n = +1$ when n is even, and -1 when n is odd.

Returning to our skunk: we know that multiplying by -1 has the same meaning as "opposite of...", and that two negatives make a positive. The sentence about the skunk is just an idiotically long chain of negatives, or opposites of opposites of...

We can short-circuit this: we have only to count the number of negatives; by regarding them as (-1)× 's, we conclude that:

an **even** number implies **+1**, **(no negation)**; and

that an *odd* number amounts to a single (-1), namely **negation** *of the final statement* (in this case - flattening the skunk). The above report contained 7 negatives (check!), so we can happily conclude that the skunk is alive and well (endangered odour).

7.3 Cutting the cr...

From here on we will deal with *GEN*eralised *SHORT*cuts used for extracting required conclusions from given information.

(In the case of party political broadcasts you can simply assume the opposite, but here we deal with serious matters and therefore need more elaborate tools.)

Asking how long the (Wagner) opera will last, you are told that "after it ends you will have an hour to recuperate before the midnight meal". If you know what time the performance starts you can work out its length in your head without recourse to special methods.

However, if the answer to your question is "The very famous conductor flew to the wrong country by mistake; the performance will therefore begin two hours late; he will however conduct the music at twice the normal speed, (which does not matter, it will still feel very long); in any case it will end four hours earlier than last time, when it took twice the normal time because a new paramedic attendant (who was not briefed about Wagner) kept interrupting the performance because he thought the sopranos were having hysterical fits" - you might *still* be able to work out in your head how long the performance normally takes, but the process would be tedious in the extreme.

Besides sticking to Mozart next time, we here resort to a simple, elegant mathematical technique that cuts through these convoluted tales and produces the required result.

The reason we used opera here is that exercises which begin with the school relic "a train leaves..." make listeners run away faster than the train... Let us see: a train leaves London, travelling to Glasgow at 80 mph; one hour later another train leaves Glasgow bound for London at 100 mph; the distance from London to Glasgow is 404 miles; When the trains meet which one is nearer London?

(Are you still there?)

If you do not know the answer instantly you might have not run away, but neither did you pay attention: the 'maths cut-out' had set in. Were it about anything other than these 'algebtrains', say two alligators racing towards each other, you would not ponder much about which one is nearer *anything* once they get to chew each other's head off...

Back to the opera...

We start by simplifying the presentation of our data. First we can invent a shorthand for each entity involved in the story.

A remarkable, still very controversial, discovery has recently been made in some extra-school territories: Apparently, it *is* possible to do 'algebra' with symbols other than *X*. Indeed it is more helpful to use letters that remind of what they represent. (Here the nearest we get to 'algebra's front-end symbol' is only the multiplication sign ×.)

So, for "**D**uration of the **o**pera **p**erformance" we can write 'Dop', or just *D*.

However, in this tale there are three different durations: the normal one, the previous one which was twice as long as the normal, and today's halved one, so we must decide which one we are going to label *D*. Let us use *D* for the *normal* duration (although everything would work equally well if we chose any of the others).

Similarly, we can abbreviate all the other elements that are involved: "today's ending time": *TE*; "previous ending time": *PE*. Since we relate ending times to the opera's duration, the starting time obviously matters too. Since we are dealing with German opera, we may assume that the performances' *normal* starting time is the same every day. We label it *ST*. (The *normal duration*, as deduced from this tale, does not actually depend on what part of the day all this happens, so we would expect that at some point this *ST* will 'drop out' of the proceedings.)

Next we must decide what units we are going to use for presenting the data. All the timings must, of course, be stated in the same unit – we choose 'hour', so we know that '*D*' means '*D* hours' and '½' means '½ hour'. (It should be born in mind that for any quantity*, say '*n*', you must be prepared to answer the question "*n* what?")

The tale is further abbreviated by using symbols for the relationships between quantities: 'later than', 'longer than', 'after' etc. can be expressed by '+', and 'shorter than', 'before' etc. by '-' . Thus "four hours before the previous ending time" becomes simply *PE*-4.

'Of', as in 'half of..' and 'as', as in 'twice as...' we know means 'times', so is written '×'.

7.4 Why, suddenly, is there no life without 'equations'?

Now we come to the key point: nothing can be worked out without *information*.

Anything that is regarded as *information* says something about something, and this cannot be done without using the word '*is*' or its equivalents: 'was', 'be', 'has been', etc.. Where this word is absent, it is nevertheless *implied*, e.g.: " Beethoven asks to scrap old instruments" can be re-phrased as "Beethoven's view about old instruments ***is*** that they should be scrapped". Admittedly '*is*' is quite short, but we use the even shorter symbol '=' .

So, all condensed renderings of information contain the *equal* sign (=) and are called *equations*.

It should of course be 'equalions'. Some time in the past a careless printer failed to clean his plates, and a speck appeared across the 'l' and so 'equations' came about... (may be).

Armed with these abbreviations, we can now condense the six lines of information into something one inch long.

---

* Unless it is a numerical multiplier or divider (like 'twice' or 'quarter'), or a ratio between identical entities.

We go through the story summarising the information step by step, beginning with today's events. It tells us that today's ending time ('*TE*') will be ('=') the normal starting time ('*ST*') plus ('+') 2 hours'delay ('*2*'), plus ('+') the performance time of half the normal duration ('*½ D*'):

$$TE = ST + 2 + \tfrac{1}{2} \times D \quad \text{------------------------} \quad (1)$$

The next part of the tale concerns the previous event: the previous end of the performance ('*PE*') was (=) the normal starting time ('*ST*') plus (+) a performance of twice (2×) the normal duration ('*D*'):

$$PE = ST + 2\times D \quad \text{------------------------} \quad (2)$$

Lastly, we have the information concerning the relationship between the events of both days: "Today's ending time is four hours earlier than the previous ending time", or, rephrased:

"today's ending time ('*TE*') is equal to (=) the previous ending time ('*PE*') minus (-) 4 hours:

$$TE = PE - 4 \quad \text{------------------------} \quad (3)$$

(N.B. All we have done so far is to unravel a written passage and record its relevant information in a concise way.)

We must never lose sight of what we are after; in this case it is the normal duration, D, not PE, not TE or ST, just D. To determine the value of D, as we will soon find out, one needs a *single* equation that contains the D and otherwise only numbers i.e. no other unknown quantities.

This is easier done that said. We can start with any of the three equations, the choice will not affect the final result, only the length of the process. It turns out as most convenient to start with the third equation. Although we are not interested in the *TE* and *PE* we have to use the information contained in this equation because the required result can only be achieved if *all* the given information is used. What we do is to replace these unwanted *TE* and *PE* by something of equal value and which contains D -the item that we *are* interested in. To do this we now use the information we have in the remaining two equations:

The first one, $TE = ST + 2 + \tfrac{1}{2}D$, tells us that TE can be replaced by $ST + 2 + \tfrac{1}{2}D$.

The second, $PE = ST + 2D$, tells us that PE can be replaced by $ST + 2D$.

So, we take the third equation $\qquad TE = PE - 4$

and perform these replacements: $\qquad ST + 2 + \tfrac{1}{2}D = ST + 2D - 4$

If two things reach the same height while both stand on top of ST, they are also equally tall by themselves; that is why ST can, as we expected, be dropped (from *both* sides)*, leaving us with what we sought, a single equation containing D, and otherwise only numbers:

$$2 + \tfrac{1}{2}D = 2D - 4$$

Nice, compact, *and*: useful, as we will see.

---

* This was crucial for another reason: with three equations i.e. three information items, we can only find the values of three unknowns (here TE, ST and D)

Now we take another respite from the opera and perform the same process with the inescapable "A train leaves..."

The question was "A train leaves London travelling to Glasgow at 80 mph; one hour later another train leaves Glasgow bound for London at 100 mph; The distance from London to Glasgow is 404 miles; when the trains meet which one is nearer London?"

This time let us be serious, and change the question to "*when* do the trains meet?" i.e. the time that elapses between the departure of the first train and the collision.

We adopt the same procedure as for the opera: we summarise the relevant elements and use shorthand whenever possible:

- Distance L - Glw: 404 M.

- Two speeds: train 1 (L-Glw) 80 mph; train 2 (Glw-L) 100 mph.

- Elements implied by 'meeting': "Meeting" (apart from the lie used by secretaries of people who don't want to come to the phone) means "being in the same place at the same time". To express this we need the travel times of the two trains, T_1 and T_2, and distances covered before meeting, D_1 and D_2.

As the unit of distance we will use 'mile' because the L - Glw distance was given in miles, and as the unit of time we will use 'hour' because the speeds where already given in miles per hour.

We can now use all the given information to connect these elements in the form of *equations* - love without rows is not a marriage, and a sequence of items without '=' is not information.

Thus we have the following data:

The combined distance covered by the two trains until they meet is the distance between London and Glasgow, i.e.

$$D_1 + D_2 = 404 \quad ......(1)$$

Travelling times T_1 and T_2 come to an end together, but T_2 started 1 (hour) after the start of T_1, therefore:

$$T_2 = T_1 - 1 \quad .....(2)$$

Then we have data on speed, but we must first define 'speed':

If you travel at 30 miles per hour for 1 hour you cover 30 miles

If you travel at S miles per hour for 1 hour you cover S miles

If you travel at S miles per hour for H hours you cover H×S miles

Train 1 travels at 80 mph for T_1 hours, so the distance it covers (D_1) is:

$$D_1 = T_1 \times 80 \quad ....(3)$$

Train 2 travels at 100 mph for T_2 hours, so the distance it covers (D_2) is:

$$D_2 = T_2 \times 100 \quad ......(4)$$

We now have four items of information; this is exactly what is needed to determine the values of the four elements T_1, T_2, D_1 and D_2.

However it is only one value that we are ultimately interested in: T_1 (the time that elapsed between the departure of train 1 and the meeting). Therefore we proceed to replace the other elements with substitutes of equal value, and which, eventually, will contain the desired 'T_1', accompanied only by (known) numbers. We do so by 'using up' the items of information, i.e. the equations. In some situations this is not easy, but in this case it is very easy, even fun...

The order in which we do this does not matter, so let us begin with the 1st one,

$$D_1 + D_2 = 404$$

We need an item of information that can replace D_1 with something containing

T_1: the 3rd equation, $D_1 = T_1 \times 80$, enables us to replace D_1 with $T_1 \times 80$:

$$T_1 \times 80 + D_2 = 404$$

No equation says that D_2 can be replaced by something containing T_1; only the 4th equation, $D_2 = T_2 \times 100$, says anything about D_2. We must use this information at some point anyway, so we use this equation now to replace D_2 with $T_2 \times 100$, hoping that another item of information will eventually replace T_2 with T_1. Thus the, now further 'revised', equation 1 becomes

$$T_1 \times 80 + T_2 \times 100 = 404$$

Which equation can rid us of T_2? Only one is left, the 2nd equation $T_2 = T_1 - 1$, and this one also happily enables us to replace T_2 with something that contains what we are interested in, the T_1, i.e. T_2 is replaced by $T_1 - 1$. And so, we end up with:

$$T_1 \times 80 + (T_1 - 1) \times 100 = 404$$

Again, an equation containing only one unknown, the desired T_1, and otherwise only numbers.

7.5 From a short story to a short answer

Rendering information into compact equations is not done just to save paper. Our purpose in engaging with all this information was to determine the normal duration (D) of an opera performance and how long a train travelled before compacting another. The form into which we summarised the information, the equation, lends itself to application of a *GEN*eral *SHORT*cut for extracting the answers. This is what is known as "solving the equation". In our cases, the first exercise (opera) would be "solving for D", the second (trains) "solving for T_1". This method is termed '*GEN*eral' because it works in the same way no matter what the story is about (or what letter we use for 'time span').

Before we go further into methods of solving equations a few points must be made about the nature of equations, in order to ensure that they *can* be solved, and thus yield the results we want. Not everything that *looks* like an equation actually carries useful information, just as not everything that *sounds* like a statement actually means anything.

I. A single equation must contain only one element whose value is unknown, that is, the element that is to be evaluated from the equation.

This principle can be compared to certain forms of verbal information: let us suppose you want to know how old the milkman is. You are told he is twice as old as you. Assuming you can still remember your age (otherwise the milkman would be long dead) this statement contains only one unknown (the milkman's age), so you can readily work out its value. If, however, the answer were that he is 22 years older than his son, you are none the wiser (unless you are the son (and know so)).

When there *is* more than one unknown more information, i.e. more equations, is needed. In such cases as we have seen above, the several equations with several unknowns are combined into a single equation with a single unknown.

II. The information provided by the equation must not be something that is true *irrespective* of the unknown's value; naturally such information would be useless in establishing any one particular value for the unknown.

For example, '$g + 8 = g + 8$' looks like an equation, but is about as informative as the deep insight of "a man's got to do what a man's got to do".

A less obvious example might be $(W + 3)^2 = W^2 + 6W + 9$. This tells us nothing about any *particular* W, because, as we have already seen, the right side of the equation is just another way of writing the left side, in this case without the brackets. Thus the equality is true whatever W is.

This too has a commonplace analogy: sentences that pretend to describe or explain something, but only repeat it in different words, e.g. "the party-spokesman's speech was all political" (what else…)

III. The information must be *possible*. Not everything that looks like an equation contains information that is possible, (again, a common phenomenon: not all things that sound like a sentence are plausible, e.g. "the tax inspector told me a joke"...)

'$A + 5 = A - 2$' is not possible: no number exists which, if increased, by 5 or by anything else, will yield the same value as when it is decreased.

The beauty of maths is that when nonsense is rendered into mathematical *SHORT*hand, it is shown up much more readily than in verbal confabulations.

7.6 Not 'how to', but *HOW* to 'how to'

While we will now find out how to extract the value of the unknown in the equation, in other words to 'solve the equation', our *main* aim really is to *deduce the method* for doing so.

Although we already have two equations that cannot wait to be solved, it will be more convenient to start with the simpler example of:

$W \times 3 + 2 = 8$. ('W' is short for What We Want to knoW).

We should never seek solutions until we have first defined exactly what the problem is. Our problem here is that while the equation tells us how much '$3 \times W + 2$' is, what we want to know is what 'W' is, on its own. Maths or no maths, the principle is quite simple: we need to strip everything other than W from the side of the equation containing the W, so that we end up with the equivalent of 'W is…', namely: '$W = ...$'. This will of course only be useful if the other side of the equation consists only of things the value of which we know, for instance just numbers.

Here we need to kill the '× 3' and '+ 2', but how? Erase them? Shoot them? We must not forget that our only hope of finding out what 'W' is lies in the information we were given about it. This information is summarised in the form of an equation (*equation*), which retains the information only as long as the two sides remain *equal* to each other. So whatever changes we make, we must maintain this equality. If we simply pillage the left side it will no longer equal the right side, and the information is destroyed.

We *must* change the left side, from **W×3 + 2** to **W**. Some change must therefore also be made on the right side. The methods we choose for removing the unwanted elements from the left side must make clear the change that should be made on the right side to ensure equality after the changes.

The general tactic is to "do the same thing" on both sides. But if, for example, we 'shoot' the '+2' on the left, it is not quite clear whom to shoot on the right so that the killing fields emerge equal. We need a 'killer of +2' which will, if let loose on the other side too, inflict equal damage there. What other methods can neutralise a plus ?...

A *minus*. Setting off a '- 2' on the left side will annihilate the '+ 2', leaving W × 3, and *this* can also be done on the right side, to the **8**: It takes 2 away from it, leaving 6.

We can be sure that the sacred equality has been maintained: if two equal things suffer the same loss, they remain mutually equal.

At this point the reader can be liberated from one of these greatly misleading nonsenses that are routinely used to explain this topic. In our example, we began with

$$W \times 3 + 2 = 8,$$

and following the first part of the 'clean up' of W we got W×3 = 8 - 2. We keep hearing that "the +2 *moved* from one side to the other, inverting its sign on the way". Whether we call this area *Gens & Shorts* or Algebra, a haulage company it is *not* (not even one that delivers the goods upside down). **Nothing was actually *moved*.** There was a clear reason for the operation and it had nothing to do with any 'moving' even if it *'looks'* like it, just as **6** +3 = **9** does not imply that adding 3 turns things upside down... This issue is dwelled upon because maths education should not be about learning 'useful recipes for correct performance', but about the pursuit of *reason*.

We began our exercise with W×3 + 2 = 8, and have so far arrived at W×3 = 8 - 2, i.e. W×3 = 6. We have still to liberate the W from the '×3'. As before, we need something that kills '×3' and that can also inflict the same damage on the other side so that equality is maintained. In other words, we need an operation that can turn W×3 into W, i.e. neutralise a multiplication, and that is also applicable to the other side: Naturally, a *division* (by 3). So, applying this 'killer of ×3,' to both sides, namely, dividing by 3 both sides of the W×3 = 6 gives W on the left, and 2 on the right:

W = 2 i.e. the straight answer to "what is W".

Once again, if you are told that the ×3 has '*moved* from the top left to the bottom right', reclaim your tuition fees.

What is the absolute next step, even if both the phone *and* doorbell ring? One *checks* the result by going back to the original equation, in this case to W×3 + 2 = 8, replacing the W by the 2, and verifying that the two sides still balance.

Students whose curiosity and courage have not been suppressed, will ask why '+2' was removed before '×3' rather than the other way round. Not only is this a good question, but worrying about such questions is more important than knowing how to solve equations! Our *main* aim here is not learning the methods for solving equations, but learning how to *develop* the methods on one's *own*. In the process of doing so, one must certainly address the question of whether the order in which two given steps are done matters, and if so, what the order should be.

The obvious way of finding this out when left to one's own devices is to try both ways and see what happens.

So let us do the above in the alternative order, namely, start by killing '×3' (by dividing by 3).

Before we start dealing with each side in turn, we must remember that it is the *entire side,* not just some of the elements, that is kept in balance with the other side. So, whatever changes we make must be made to the whole side. If a side consists of $6 + 21\times K + H$, and we add 4, there is no difference between adding it to the H and adding it to the whole $6 + 21\times K + H$. The same for subtraction. However, if we need to multiply the side by 3, and write $6 + 21\times K + H\times 3$, it is only H that is multiplied, not the whole side. To indicate the whole side is multiplied, we must write

$$(6 + 21\times K + H) \times 3$$

which we know means

$$6 \times 3 + 21\times K \times 3 + H \times 3$$

Remember: $(3 + 4)\times 5$ is $3\times 5 + 4\times 5$, whereas $(3 \times 4)\times 5$ is just $3 \times 4 \times 5$.

The same applies to division. Let us say we have $6 + 21 = 9 + 18$. If we want to divide 6 by 3, and want both sides to remain equal, we must divide *every* component (on *both* sides) by 3. Try it.

In our $W\times 3 + 2$ (= 8), if we want to start by killing off the '×3' by dividing by 3, we must do it to the *whole side* $W\times 3 + 2$, i.e. dividing *both* the $W\times 3$ *as well as* the 2: $W\times 3 / 3 + 2/3$ (which is $W + 2/3$). Then, of course, we do the same (division by 3) on the other side: 8/3. So we get

$$W + 2/3 = 8/3.$$

Next we must rid the W of the '+ 2/3' by subtracting 2/3 from it, and, of course, from the other side too. This now gives us

$$W = 8/3 - 2/3.$$

The fractions have a common divider (denominator) so we can write this as

$$= \frac{8-2}{3} = \frac{6}{3} \; (=2)$$

Thus we end up with the same result, W = 2, as we did when we began by eliminating '+2'. The conclusion is that both methods work, but the first one was a little shorter.

In this case the difference was small, because the whole thing was small, but in real life it does make a difference. We therefore deduce that it is best to begin by dealing with + & -, and then turn to × & /.

Our next case is:

$$3\times W - 2 = W + 4$$

If we cleanse the left side of '-2' by throwing '+ 2' at it, and do the same at the other side we get

$$3\times W = W + 4 + 2$$

which is $3\times W = W + 6$

then if we kill off '3×' by dividing it, and everything else, on both sides, by 3, we end up with

$$W = W/3 + 6/3$$

or $W = W/3 + 2$

and we have a nice clean lone W in front of '=' waiting to be told by the other side how much it is worth. But *this* time it is like "a man's got to do one third of what a man's got to do, and then go on for a couple more"... Since W is unknown, it would be helpful *not* to use it as part of the answer to "what is W?"

The conclusion is that before we do anything to isolate the W on one side, we must remove any trace of it on the other side, the place that will hold the answer to "what is W ".

So, we need go back to the

$$3W - 2 = W + 4$$

It does not matter in principle which side we choose W to remain on, but in practice one of the sides is usually found to be more convenient. In this case it is the left side, so we will clear the right side of W. Again we use our *"Terminator 9"* method: fire a '- W' at it, leaving behind 4. No more Ws there. To uphold equality the same ruthless violence is then meted out to the other side too (a common form of justice), so we fire '-W' also at the left side, leaving

$$3W - 2 - W = 4$$

Next we must bear in mind that our purpose is to end up with "W is ...", i.e. a phrase that contains W only once. So we need to rephrase the left side in such a way that W only appears once.

In our case this is quite simple: 3W - W is just 2W, so the left side becomes

$2W - 2 = 4$ (while the right side is resting, because the events on the left did not change its value).

In a more general case where the left side were, say, $3\times W + n\times W$

We also know a way to rewrite it using one W only: $(3 + n) \times W$.

[Thereafter, to isolate the W, one divides by $(3+n)$].

We know the rest: the '- 2' on the left is eliminated by '+2', which is also applied to the right, giving:

$$2W = 4 + 2 \quad (=6).$$

Finally, we rid 2W of the multiplying 2, using a '/2', and do the same to the 6 on the right, (6/2 i.e. 3) ending up with

$$W = 3.$$

And - we have not forgotten that nothing happens now until we have checked that the two sides of the given equation $3W - 2 = W + 4$ are indeed equal if W is 3.

The *GEN*eral *SHORT*cuts that lead from an equation, which is a distilled rendering of given information, to the answer, can thus be listed as follows:

1. Eject all unknowns from the 'answer side'.

2. Rephrase the side containing the unknown in such a way the unknown appears only once.

3. Step by step rid this side of anything other than the unknown, using operations that also leave meaningful results when applied equally to the other side.

Now, at last, we are ready to find out how long the opera takes, and when the trains meet!

The contorted 6-line tale on the opera boiled down to

$$2 + D/2 = 2D - 4$$ (one inch long, as promised).

There are Ds on both sides, so we must start by clearing one side of them.

If we chose to clear D from the right, by subtracting 2D from it, and do the same on the left, the left side would become $2 + D/2 - 2D$, that is, 2+ half D less 2D, which results in a negative quantity of D: $-1\frac{1}{2}$ D. This is not convenient, because we would then have to get rid of '-' by multiplying it (and everything else) by -1.

We can avoid all this by choosing to clear D from the *left* side: Subtracting $D/2$ from both sides yields (as stage I above)

$$2 = 2D - D/2 - 4$$

Now we want only one D (on the right), which is easily achieved by writing $2D - D/2$,

with D used only once (by 'taking D outside a bracket'): $(2 - 1/2) \times D$, which is 1.5D.

So in stage II we have

$$2 = 1.5D - 4$$

Stage III entails clearing the side containing D, now on the right, of anything other than D: the - 4 is killed by '+4'. Restoring the balance by doing the same on other side, inflates the 2 to 6:

$$6 = 1.5D$$

Finally the multiplier 1.5 is obliterated by dividing by 1.5, which, applied also on the left, divides the 6 $(6 / 1.5)$, leaving 4:

$$4 = D$$

In this way, given the convoluted information on the tribulations of performing one of Wagner's shorter operas, we are able to conclude that the normal duration of the opera is 4 hours.

(We know that '4' here means 4 *hours* because, firstly, it must be 4 *something*; secondly, D is to do with *time,* and thirdly, the unit of *hour* was set from the start, in the story, and must be remembered to the end.)

But there will be no midnight meal (remember?) before verifying that the two sides of the given equation $2 + D/2 = 2D - 4$ are equal when the solution, the 4, is substituted for D.

Information on the London - Glasgow trains was summarised as:

$$T \times 80 \quad + (T\text{-}1) \times 100 = 404$$

One of the Ts is enclosed in brackets; since we want to isolate T, we must first get it out of there. We know how to write $(T\text{-}1) \times \mathit{100}$ without brackets: $T\times \mathit{100}\ - 1\times \mathit{100}$,

so, liberated from brackets, the equation becomes:

$$T\times 80 \quad + T\times 100 - 100 = 404$$

This, with T written only once $[80T + 100T = 180T]$ is

$$180T \quad - 100 \; = 404$$

To strip the T bare, we first oust '- 100' by hitting it with +100, which also affects 404 on the other side to give:

$$180T \qquad = 504$$

Finally '$180\times$ T' is reduced to $1\times$ T by dividing it by 180, and with the 504, too, divided by 180 we end up with:

$$T \qquad = 504/180$$

$$= 2.8$$

2.8 what? Hours, as was decided at the outset (to use hours & miles).

All we have to do to put this in a more familiar form is work out what '.8' of an hour is: 0.8 is 8 tenths; 1tenth of an hour is 6 minutes, so 8 tenths of an hour is 8×6 minutes.

The trains met 2 hours and 48 minutes after the first one left London.

7.7 Many questions with many stories

Given two unknowns, such as Gug's age and Mike's age, and only one item of information, such as "their ages add up to 60", we are left with the possibility that Mike is sixty and Gug has only just entered the world, or that they are thirty year old twins, and many other possibilities besides. If, however, we have another piece of information, such as that Gug is twice as old as Mike, their ages are easily determined: forty and twenty.

The above we could in our head, but let us take as a similar example **3W + 2U = 17.** Again, the single item of information, the single equation, is not enough for finding the value of two unknowns (W and U). If, however, we ask nicely and are told also that $U\text{-}W = 1$, we can work out what both W and U are. This always involves proceeding first towards a single equation with a single unknown. But how?

If, instead of $U - W = 1$, the second equation were, for instance, $W = U - 6$, this would be easy, as in the trains' case*: the second equation tells us explicitly what W is, rather than some *indirect* information about W, as in $U\text{-}W=1$. We could therefore rid the first equation of one of the two unknowns, the W, by putting this (U-6) in its place:

$$3\,W \quad + 2U \; = 17$$

$$3(U\text{-}6) + 2U = 17 \qquad \text{- only one unknown, the U.}$$

How to solve this we now: (1) know; (2) should do; (3) do again if we do not get U = 7.

---

* Where, in fact, we had not two but four equations and four unknowns.

But what do we do with a pair like $3W + 2U = 17$ and $U - W = 1$?

(The usual name for this topic is *"simultaneous equations"*).

There are two methods. We will use the one which is *usually* more convenient when there are only few simultaneous equations (with equal number of unknowns, of course). The other method, which we will not go into, is used mainly in cases of a large number of simultaneous *linear* equations (equations, like so far here, where no unknown is multiplied by another or by itself), especially where computers are used to solve them.

As we have seen, we could proceed easily if at least one of the equations had one of the unknowns posted alone in front of the '='. This could then be used to replace the same unknown in the other equation, leaving the latter with only the second unknown. So, again we employ our problem-solving method: convert the *given* case into the form which we *know* how to handle.

We start by converting one of the two equations into the form 'W = ' or 'U = '. A glance at these two equations suggests that this is most easily done with the $U - W = 1$: isolating U by ridding it of the -W. This is done, naturally, by applying the '+W antidote', and, of course, safeguarding the *'equation'* by doing the same on the other side. Thus this equation becomes

$$U = 1 + W$$

From here on it is all downhill. In

$$3W + 2U = 17$$

we replace U by the $1+W$ and get

$$3W + 2\times(1+W) = 17$$

which is one equation containing one unknown, with only two things left to do: solve it.

'opening' the brackets: $3W + 2 + 2W = 17$

wobbling your hand only once,

namely, writing it with a single W: $5W + 2 = 17$

baring the W: $5W = 17 - 2$ $(=15)$;

and $W = 15/5$

(= 3)

Having found the value of one unknown, the W, we can now use either of the two equations to find the value of the other unknown, the U. The easiest one to use is of course the simpler of the two, the $U - W = 1$, especially since it has meanwhile been converted to the form "U = ... ", that is, $U = 1 + W$; (W was found to be 3, so $U = 4$).

We said *two* things to do: Even with the house on fire, the parrot and you escape only after checking these results in the other equation - whether $3W + 2U$ does indeed equal 17 if $W = 3$ and $U = 4$.

7.8 A square deal

What do we do with an equation looking like $L^2 = 9$?

Basically the same as we have done before. We aim to strip the L-side down to "L = ". In this case it is just the little ^{2} up there that needs to go. The only question is what procedure can not only reduce L^2 to L but also leave a meaningful result when applied to the other side, thus ensuring that equality is maintained.

L^2 is the result of the multiplication of two Ls, in other words L is the *2-way fragment*[*] of L^2. So, 'taking the 2-way fragment' on the left of $L^2 = 9$ will yield the desired L there, and the same must of course be done on the right: taking the 2-way fragment of the 9, which we write $9^{1/2}$.

(The confusionists, remember, call this 'taking the square root' of 9, and write it $\sqrt[2]{9}$).

Thus if $L^2 = 9$, then $L = 9^{1/2}$. The 2-way fragment of 9 is 3 (because $3 \times 3 = 9$), but the 3 is not alone in this role: we already know that also $-3 \times -3 = (+)9$

So, given the equation $L^2 = 9$, the solution is $L = 9^{1/2}$

$= 3$ *and* -3, or, in short, '± 3'

Thus we have solved an equation containing a *selfmult*. So far so good, but what about something like $3L^2 + 5L = 8$, that is, an equation containing both L^2 and L? How can this be rewritten using only one L (which will eventually stand alone on front of '=')?

We know now that somehow we will have to involve taking 2-way fragments*, the tool which dismantles the 'squaring', i.e. gets rid of the ^{2} in L^2.

The problem, as we have seen before, is that treatments can be administered only to *entire sides.*

If L^2 is the only occupant of the side there is no problem, we simply do $(L^2)^{1/2}$ (or the ugly $\sqrt[2]{L^2}$) which yields the desired L. The question now is whether $(...)^{1/2}$ can be applied to a whole side such as $3L^2 + 5L$.

$(...)^{1/2}$ can indeed be applied to a *group* of elements if they are *multiplied* together, because the

2-way fragment of $(A \times B)$ is equal to

(2-way fragment of A) x (2-way fragment of B)

i.e. $$(A \times B)^{1/2} = A^{1/2} \times B^{1/2}$$

We can (should) try this with numbers: $(4 \times 25)^{1/2}$ is $100^{1/2}$, which equals 10,

this, 'opened up': $4^{1/2} \times 25^{1/2}$ is 2×5, also gives full marks.

Another example: $(9 \times L^2)^{1/2}$ is $9^{1/2} \times (L^2)^{1/2}$, which is $3 \times L$.

---

[*] Ex. 'square root'.

Note that trying out with numbers is not exactly a mathematical proof. It might work with the numbers you try but not with all others. Still, it gives a 'feel' of what is going on, and the more examples one tries the better the feel...

In attempting to solve the equation $3L^2 + 5L = 8$, however, the problem is that $(...)^{1/2}$ is applied to a *sum*, in this case $3L^2 + 5L$,. but a fragment of a *sum*, is not equal to the sum of the fragments: e.g. $(A + B)^{1/2}$ is *not* the equal to $A^{1/2} + B^{1/2}$.

Again, try it with numbers:

$$(9+16)^{1/2} = 25^{1/2}, \text{ which is } 5,$$

but

$$9^{1/2} + 16^{1/2} = 3 + 4, \text{ which is.... not } 5.$$

Besides, even *if* $(A + B)^{1/2}$ *were* equal to $A^{1/2} + B^{1/2}$, or, as in this case, if $(L^2 + L)^{1/2}$ *were* equal to $(L^2)^{1/2} + L^{1/2}$, the last thing we want in our quest for a neat L is to introduce things like $L^{1/2}$ i.e. the 2-way *fragment* of L (and even less when called 'root').

Let us consider another suggestion.

Since the difficulty arises from (i) the square, and (ii) the simultaneous presence of L^2 and L's, could not both these problems be taken care of by dividing the whole side (the $3L^2 + 5L$) by L i.e. $\frac{3L^2 + 5L}{L}$? That would remove the ^{2} from L^2 ($3L^2/L = 3L$), and eliminate the L from the 5L altogether ($5L/L = 5$). This looks promising, but the other side also has to be divided by L, and this would produce $8/L$, i.e. the unknown appearing as a *divider*, which again would only make things worse.

Before showing how even an ordinary mortal can find a way around this, let us see how equations of this sort can arise in the first place.

Let us suppose that you have £8 to spend on a square framed mirror. The glass costs £3 per square metre, and the frame £1.25 per metre (of the framing material). You want to know what size mirror you can afford, in other words what greatest length / width of the mirror.

We will call the unknown length of the mirror L.

The area is therefore L^2, and the length of the frame 4L.

The total cost will therefore be $3 \times L^2 + 1.25 \times 4L$, which is $3L^2 + 5L$, and this must equal 8.

The exercise is then to find the value of L, given that $\mathbf{3L^2 + 5L = 8}$.

The formal name of this task is "solving a *quadratic* equation". ('*quadratic*': to do with *square e.g.* L^2).

This begins to raise terrifying recollections of '*THE FORMULA*", the scary sight of

$$\frac{-b \pm \sqrt{b^2 - 4ac}}{2a}.$$

Two things are fairly certain here:

- one is unlikely to reach adulthood without having had engaged with the above, and yet:

- if cars were driven only by those who, to save their life, could show exactly where this thing comes from the Green movements would have to set oilfields alight to justify their existence.

We will find out how unnecessary this FF (*Formula Fear)* is. We will show how anyone could deduce the method for solving such a 'quadratic equation', how anyone could evolve the above *'formula'* by a process which is very much like a detective story: looking for clues, coming up with logical conclusions and deciding on the most promising next steps.

Relatively few people will ever encounter quadratic equations in their daily (or nightly) life but everybody will benefit from the ability to apply orderly thought processes, such as described here, to solve newly met problems. Being made to memorize, and exercise the use of *'the formula'* is therefore less important than being taught how to *derive* it.

In case the sight of the following appears like looking down a black run ski slope, it should be born in mind that *here*, in contrast, one can take it as slowly as one wishes, (and yet reaching the bottom is every bit as exhilarating).

Rather than using the school routine of *telling* what each step *is*, we will show how to *deduce* what it should be, just using common sense. The approach we take may well be the one used by the first person to have accomplished this task.

A final note before we start: do not be put off by the length. Length and difficulty should not be confused. What matters is that each step is easy, not how many there are.

So let us go back to look at the mirror and work out its size by considering the equation

$$3L^2 + 5L = 8.$$

We already know that the difficulty stems from the presence of two items with L added together, the one with ^{2}, the other without. We already tried all the direct ways of getting rid of the ^{2} (dividing through by L; taking the 2-way fragment of the whole side), and found them unsuccessful.

What we do next is really the most straightforward way to proceed.

First we summarize the problem: we have something with L^2 and L but need something where the symbol L appears only once.

Next, what would *you* do were this not maths? Say you needed a big elephant and a little elephant but the shippers can only accommodate one animal. Naturally you would ask yourself: "can I think of a single animal which contains both the things I want?" Having asked this correct and obvious *question* it would not take you long to conclude that you need to order a *pregnant* elephant. Why not do the same thing here?

Have we ever seen something that has L written in it only once, and yet is equal to something that contains both L^2 and L ?

? ?

Yes, a couple of pages back. $(\boldsymbol{L} + s)^2$! - which equals $\boldsymbol{L}^2 + 2s\times\boldsymbol{L} + s^2$.

Note: the letters L and s are used now in two different roles. The L is the *unknown*, the one that will be established only *after* solving the equation, while s represents a quantity that *is* known before starting to solve the equation. Even though the quantity is known one uses letters (rather than numbers) whenever the story does not depend on *what* the number is. For example what we said about the benefits of using $(L + s)^2$ is true for (L + *'any number'*)^2.

Upper case letters will be used for unknowns, while the *unspecified known* quantities will be lower case.

We have found a promising beginning, but two questions need to be examined.

I. $(L + s)^2$, with only one L, appears to lead to a solution. Its expanded form $\boldsymbol{L}^2 + 2s\times\boldsymbol{L} + s^2$ is *similar* to the given $3\boldsymbol{L}^2 + 5\boldsymbol{L}$, but is not *identical* to it. Can we modify the $(L + s)^2$ so that its expanded form equates *exactly* with the left side of the given equation?

We also need to establish of course what the right side of $(L + s)^2 = \ldots$ should be such that the *entire* given equation, not just its left side, can be replaced by this solvable equivalent. (In any case, the right side is simpler to deal with because it contains only numbers, no unknowns).

II. If we *do* find a version of $(L + s)^2 = \ldots$ which replicates the given equation, are we *sure* that we can *solve* this replica equation? We know it is promising because L appears only once, and we also know how to disarm ^{2}. Still, we better verify that there are no other surprises en route.

We will start by establishing some principle features which have to be present in the modified form of $(L + s)^2 = \ldots$ so that it can represent the given equation accurately:

Every general quadratic equation contains L and L^2. What, then, distinguishes one equation from another? The three *numbers*: the one multiplying the L^2, the one multiplying the L, and one that stands alone. In the case of our $3L^2 + 5L = 8$ these numbers are 3, 5 and 8 respectively*.

All the alternative possible phrasings of a given message must contain the same number of information items. Similarly, any alternative form of representing a *quadratic equation* must have room for three numbers. Such 'controlling' numbers are called *parameters*.

Where can we accommodate three numbers in $\mathbf{(L + s)^2 = \ldots}$? Certainly, the s is one of them and the '...' on the right also looks like "room to let". We will let it to 't'.

Could we accommodate the third number, r say, by writing $(L + s + r)^2$? Not really, because s & r could be replaced by a single number that equals their sum. They would contribute nothing more than writing 3 + 11 instead of 14. So where do we put the third one? Wrong question.

---

* The information contained in the *equation* $3L^2 + 5L = 8$ is unchanged if we divide all the components on both sides by 3, the multiplier of L^2, giving $L^2 + 5/3\,L = 8/3$ i.e. only two numbers (5/3 & 8/3), and this would simplify the whole process. However, for reasons that are good but too long to explain here, general quadratic equations usually *do* appear with three numbers, and we therefore proceed with this form henceforth.

What we should ask is *how to deduce* where to put it. We must not lose sight of what we are after: trying to match the solvable replica '$(L+s)^2 = t$' with the given $\mathbf{3L^2 + 5L = 8}$.

We can see some connection between the t and the 8 on the right; we also know that when we rewrite the replica $(L+s)^2$ without the brackets (we know how) the L-component appears with some multipliers ($2\times s$) and this could be related in some way with the 5 that multiplies the L in the given equation.

But how can the replica $(L + s)^2 = t$ provide something that multiplies the L^2, like the 3 does in the given equation? It can only come about if instead of $(L + s)^2$ we had

$$(\mathbf{r}\times L + s)^2$$

because when we write this without the brackets (treating the **r**×L as if it were a single symbol) it looks like:

$$(r\times L)^2 + 2s\,(r\times L) + s^2$$

$$\downarrow$$

$$r^2\times L^2$$

so, by introducing into the $(L + s)^2 = \ldots$ something that would correspond to the 3 that multiplies the L^2 in the *given* equation, we found a place for the third number in the 'solvable replica'.

To summarise: we established that a general replica for a quadratic equation can be written as:

$$(r\times L + s)^2 = t$$

We must of course proceed to find the exact correspondence between r, s & t in the replica and the 3, 5 & 8 in the given equation (we might have already noticed a connection between r^2 and 3, the multipliers of L^2, and between $2s\times r$ and 5, the multipliers of L), but at this stage we first want to verify that there are no obstacles on the way to *solve* this $(r\times L + s)^2 = t$ equation.

Solving the equation means of course getting the L to stand alone in front of the '=', and ensuring the other side contains only r, s, t and, possibly, some other numbers.

We want to liberate the L, unpack it. It is hard to unpack things from the inside. So we start by ripping away the ^{2} from the outside: Jack the ^{2} ripper operates by

2-way fragmenting*, mercilessly. on *both* sides of $(r\times L + s)^2 = t$:

$$r\times L + s = t^{1/2}$$

From here on it is boringly familiar: kick the +s with -s (other side too):

$$r\times L = t^{1/2} - s$$

dump the ×r with /r (other side too):

$$= \frac{t^{1/2} - s}{r}$$

So, we *can* solve it. We just did.

* Superfluous reminder: the 2-way fragment ('sq. root') of $(xx)^2$ is simply xx.

We can return now to completing the transformation from the given $3L^2 + 5L = 8$

to the solvable replica $(r \times L + s)^2 = t$.

This entails finding the numerical values of r, s & t through their relation to the 3, 5, & 8.

(Once we have found the numerical values of r, s & t we insert them in $L = \dfrac{t^{1/2} - s}{r}$

and then know how big the mirror is, that is, the value of L.)

Let us recapitulate what there is on the playing field:

We are given an equation which we cannot solve: $3L^2 + 5L = 8$

but realize that there is a *solvable* 'replica' equation $(r \times L + s)^2 = t$

which, when rewritten without brackets: $r^2 \times L^2 + 2s \times r \times L + s^2 = t$

is similar to the given equation. It can be made *identical* to the given equation by finding suitable values for r, s & t, (which will then be used in the solution).

This can be done by comparing the given equation with the expanded form of the solvable replica equation and equating the multipliers of L^2, the multipliers of the L, and all the rest.

There is a small problem with 'all the rest' (the components which do not contain L): it refers of course to the 8 on the right of the given equation and the s^2 & t in the replica equation in which they sit on *different sides*. Like this it is unclear how to relate the s^2 & t to the 8. It would be more convenient if *everything* were on the left side (thus leaving a '0' on the right).

We *could* proceed equally well if we had all the 'non-L' components on the right, but we choose the first option for an unrelated reason:

There is another method for solving quadratic equations that is commonly used in schools. This method, when it works at all, does so only if one side is 0. For this, and also other reasons which cannot be explained here, it has become a convention to write equations in that way.

That other solving method is used in schools as it is easier to explain why it works. However, being in some ways a 'trick' method it does not lend itself to what *we* are trying to do, namely, *deduce* methods by using only orderly thought and previously understood principles of solving equations. Another reason for not using it: it involves trial and (mainly) error and... rarely works anyway unless the equations were specially made up by the teacher to be thus solvable. *Real* life is not so generous.

So let us return to the two equations $3L^2 + 5L = 8$ and $r^2 \times L^2 + 2s \times r \times L + s^2 = t$

and purge the right, but remember: no *'moving'*!

We get the zeros on the right by subtracting 8 from (both sides of) the given equation,

and t from " " " the replica.

We now have the given $3L^2 + 5L - 8 = 0$

and its replica $r^2 \times L^2 + 2s \times r \times L + s^2 - t = 0$ (for which we already have the solution)

There is something enjoyable about the way things unfold from here on, step by step, until finally the value of L is revealed:

1. Equating the multipliers of L^2 tells us that $r^2 = 3$. We want to know what r is, not r^2, so we fight off the ^{2} by taking the 2-way fragment (sq. root), of the other side too, and get

$$\underline{\mathbf{r}} = 3^{1/2}$$

2. Equating the multipliers of L tells us that $2s \times r = 5$.

   We already know what r is, so we can use this to find s:

   we strip it bare by dividing the left side by 2r, never depriving the other side of the same attention, and get $s = \frac{5}{2r}$;

   As we already know that $r = 3^{1/2}$ we get:

$$\underline{\mathbf{s}} = \frac{5}{2 \times 3^{1/2}}$$

3. Equating the leftovers: $s^2 - t = -8$. We know r, we know s, we want to know t too.

   Subtracting away the s^2 (from both sides) yields $-t = -s^2 - 8$, all rather negative...

   we want to know what t is, not -t. Multiplying it by -1 does it, and it does it on the other side too: $t = s^2 + 8$.

How good that we already know what s is. All we need to do is square it: $\left(\frac{5}{2 \times 3^{1/2}}\right)^2$

and send this to and so get: $\underline{\mathbf{t}} = \left(\frac{5}{2 \times 3^{1/2}}\right)^2 + 8$

Not so pretty perhaps but we *did* find the numerical values of r, s and t.

We have not forgotten what we did all this for: we want to find the value of L,

and L is $\dfrac{t^{1/2} - s}{r}$

All that is left then is to replace the

r by $3^{1/2}$ s by $\dfrac{5}{2 \times 3^{1/2}}$ t by $\left(\dfrac{5}{2 \times 3^{1/2}}\right)^2 + 8$

and here we go:

$$L = \frac{\left[\left(\dfrac{5}{2 \times 3^{1/2}}\right)^2 + 8\right]^{1/2} - \dfrac{5}{2 \times 3^{1/2}}}{3^{1/2}}$$

This is not as bad as it looks. Requires only patience, and for now: a calculator. Remember that brackets mean: "me first", and so inner brackets mean "the *first* among the first". Also: the nice notation $(\ )^{1/2}$ has not yet been incorporated on calculators and still looks there like $\sqrt{\ }$.

If the patience and the batteries last, L will turn out to be no more complicated than 1.

We do *not* need the calculator to check that the two sides of the given equation $3L^2 + 5L = 8$ are indeed equal if L is 1.

Unfortunately, this is not quite all. Just as $4^{1/2} = +2$ *and* -2, the above $[\ldots]^{1/2}$ too can be both + & -. In fact it should have been written up there as $\pm\ [\ldots]^{1/2}$. So, while the calculator is still warm, we do the above again, this time *subtracting* the value of the $[\ldots]^{1/2}$ (rather than adding it).

Here the calculator should come up with -2.666...(- 8/3). (Check this too, of course).

The bottom line: the length (and width) of the mirror is 1m, *but it can also be -2.666.* What does this mean?

Maths tells us that *if* someone were to cut your materials to this negative length, the mirror would still cost you £8. That no one knows how to cut negative lengths is not maths' problem. (But there *is* something more to be said about this, something really nice. See dessert)

Quadratic equations generally have two solutions. This arises from the 2-way fragment that is always involved and which has two values, one positive and one negative. (This does not necessarily cause one of the *solutions* to be negative.) Often *both* solutions are meaningful. For instance a quadratic equation needs to be solved in order to determine when a tossed pancake passes your nose; there are two solutions: one on the way up and one on the way down. The first solution derives from the negative fragment, the second from the positive fragment.

There is also the possibility of a quadratic equation having no solutions at all: this happens when the content of $[\ldots]$ is negative, because a negative number, as we found in the previous chapter, has no normal 2-way fragments ('square roots'). We will find an example of such a case below.

Finally, as all this was so much fun we will now do it again, this time not with a particular equation, that is, one with specific numbers, but with a 'generalised' equation, in which in place of specific numbers we have letters, a, b & c, representing 'any numbers':

$$a\times L^2 + b\times L + c = 0.$$

We can not solve it directly, but

will match it up with

$$r^2\times L^2 + 2s\times r\times L + s^2 - t = 0$$

-the expanded form of $(r\times L + s)^2 = t$ which we *can* solve, namely,

$$L = \frac{t^{1/2} - s}{r}.$$

To utilise this solution we need the values of r, s & t; What is *given* is a, b & c. We therefore need to express the required r, s & t in terms of the given a, b & c. We do so using the identity between the above two equations which means that:

I. the multipliers of the L^2 must be equal: $r^2 = a$. Isolating the required r: $\mathbf{r} = a^{1/2}$

II. the multipliers of the L must be equal: $2s\times r = b$.

Isolating the required s: $s = \frac{b}{2\times r}$, but

only a,b,c are allowed on the right, so we replace the r by the $a^{1/2}$

yielding $\mathbf{s} = \frac{b}{2\times a^{1/2}}$

III. the remaining items must be equal, i.e. $s^2 - t = c$.

Isolating the required t: $-t = -s^2 + c$.

Changing signs: $t = +s^2 - c$

Replacing the unwanted s by $\frac{b}{2\times a^{1/2}}$: $\mathbf{t} = \left(\frac{b}{2\times a^{1/2}}\right)^2 - c$

Now we can replace the r, s & t in the solution $L = \frac{t^{1/2} - s}{r}$ using

that is, using only the given a, b & c:

$$L = \frac{\pm\left[\left(\frac{b}{2\times a^{1/2}}\right)^2 - c\right]^{1/2} - \frac{b}{2\times a^{1/2}}}{a^{1/2}}$$

Your trust in the proof-reader is highly appreciated, but still - check the above for misprints.

So we found the solution to any quadratic equation: plug in the values of any given a, b & c, do some arithmetic and you have the result. In fact, as a bonus for all this work: *two* results, due to the $\pm[\quad]^{1/2}$. For example, if you use our 'mirror numbers' 3, 5 & -8 you get precisely the same expression as on the previous page. Note that the (+)c was -8 in the 'mirror equation' written with 0 on the right. Thus the **-c** in the $[\quad]^{1/2}$ means -(-8), i.e. +8.

Here is something nice, a devastating pitfall with an elegant way out. If 'a' is given as a negative number the above 'machine' seizes up instantly with smoke coming out of all the $a^{1/2}$'s there, (we know negative numbers can not have a 2-way fragment). So? Say the equation is $-3L^2 + 5L - 8 = 0$. We must repair the -3 problem, and we know how: by multiplying it with -1, but we must not tamper with the given information. So, we simply multiply *everything* by -1, and get $(+)3L^2 - 5L + 8 = 0$.

A *real* problem arises if the whole big $[\quad]^{1/2}$ turns out negative (and so has no real 2-way fragment). This usually means that the equation (remember, an equation is *information*) describes an unrealistic situation. For example, consider a variation of the mirror story: You *sell* framed mirrors and set the price by the length of the frame material. The glass costs you '*a*' (£) per square meter, and you receive £1.25 for each meter of the frame. You want to work out what size mirror will earn you £8 net. All the components of the equation are about 'getting', so the *cost* 'a' means *getting* '- a' (per sq. meter). The equation would then be:

$$-a\times L^2 + 5L = 8 \qquad (1.25\times 4\,L = 5L)$$

rewritten with 0 on the right: $-a\times L^2 + 5L - 8 = 0$

rectifying the '-a' (everything × -1): $a\times L^2 - 5L + 8 = 0$

This is the beauty of maths: with these numbers (b = -5, c = 8), the big $[\quad]^{1/2}$ will be negative if 'a' is more than 0.78125 and this means that unless the glass costs less than £0.78125 / m^2 you will *never* make 8 pounds, even if you frame everything from a postage stamp to the whole of New Mexico.

As to the procedure in this section, do not let its length put you off. All that matters is the revelation that each and every step can actually be *completely comprehensible*. In fact, reading it a second time it will appear half as long.

We could end here, our expression $L = \dfrac{\pm\left[\left(\dfrac{b}{2\times a^{1/2}}\right)^2 - c\right]^{1/2} - \dfrac{b}{2\times a^{1/2}}}{a^{1/2}}$ delivers the goods. But alas, it does not look like that (in)famous 'formula'..

Getting there from here involves some 'juggling' which at first sight might appear daunting, but to anyone with a little practice in maths, such things are as easy as a walk taken by a postman on strike.

To simplify matters we want to get rid of the 'two level' dividing rungs. For this we need to get the big $[(\ )^2 - c]^{1/2}$ into a form of a fraction and get it to have a divider common to that of the $\dfrac{b}{2\times a^{1/2}}$.

Remember: contrary to $\left(\frac{a\times b}{c\times d}\right)^2 = \frac{a^2\times b^2}{c^2\times d^2}$ and $\left(\frac{a\times b}{c\times d}\right)^{1/2} = \frac{a^{1/2}\times b^{1/2}}{c^{1/2}\times d^{1/2}}$, squares and fragments of *sums/ differences,* are nothing simple, e.g. $(a-b)^{1/2}$ is nothing simple like $a^{1/2} - b^{1/2}$.

Therefore, to do anything with the big $[(\ )^2 - c]^{1/2}$ in $\dfrac{\pm\left[\left(\frac{b}{2\times a^{1/2}}\right)^2 - c\right]^{1/2} - \frac{b}{2\times a^{1/2}}}{a^{1/2}}$

we must first transform the $(\ )^2 - c$ inside it to something other than a difference .

For this we must first get rid of the brackets of $(\ )^2$ (by squaring each item inside separately and noting that $(a^{1/2})^2 = a$, (and $2^2 = 4$), so $\left(\frac{b}{2\times a^{1/2}}\right)^2$ becomes $\frac{b^2}{4\times a}$.

With the $(\ )^2 = \frac{b^2}{4a}$

the $(\ )^2 - c$ becomes $\frac{b^2}{4a} - c = \frac{b^2}{4a} - \frac{c}{1} = \frac{b^2 - 4a\cdot c}{4a}$ (on a common divider).

Having changed the inside of the $[\cdots]^{1/2}$ from the

difference $(\ )^2 - c$ to the fraction $\frac{b^2 - 4a\cdot c}{4a}$ we can now 'unpack' the

$$\left[\frac{b^2 - 4a\cdot c}{4a}\right]^{1/2} \text{ into } \frac{[b^2 - 4a\cdot c]^{1/2}}{[4a]^{1/2}} = \frac{[b^2 - 4a\cdot c]^{1/2}}{4^{1/2}\times a^{1/2}} = \frac{[b^2 - 4a\cdot c]^{1/2}}{2\times a^{1/2}}$$

- all this work to get the $[(\ )^2 - c]^{1/2}$ into a fraction with the same divider as in .

With this our L = becomes $\dfrac{\pm\frac{(b^2-4a\cdot c)^{1/2}}{2\times a^{1/2}} - \frac{b}{2\times a^{1/2}}}{a^{1/2}}$

Mounting the top on the common divider): $= \dfrac{\frac{\pm(b^2-4a\cdot c)^{1/2} - b}{2\times a^{1/2}}}{a^{1/2}}$

So, finally we can carry out the desired 'ladder collapsing' $= \dfrac{\pm(b^2 - 4ac)^{1/2} - b}{2\times a^{1/2}\times a^{1/2}}$.

If we now note that $a^{1/2}\times a^{1/2} = a$, reverse the order at the top and replace the pretty $(\ldots)^{1/2}$ by the ugly $\sqrt{\ }$

we too get $\quad L = \dfrac{-b\pm\sqrt{b^2-4ac}}{2a}$.

(You still get your 1m mirror here. Check.)

7.9 A meaning-rich dessert

The negative-length mirror.

Again, the beauty of maths. The result -2.666 that maths came up with can have a logical meaning attached to it:

(We are back to the case of *paying* for the mirror.)

The square of a negative number, here the -2.666, is the same as that of a positive 2.666, so anything to do with the length squared (namely, the area of the glass), is unaffected. So, you still pay £3 for each of the $(-2.666)^2 = +7.111$ square metres, £21.333 in total.

But what about the $4 \times (-2.666) = -10.666$ meters of frame material?

For 1 metre you pay £1.25, for 2 meters: 2×1.25 etc. so your outlay for the frame material is £1.25 × its length; if the total length is -10.666 the outlay is £1.25 × (-10.666), that is: £ -13.333; *A negative outlay means a receipt.* Such would be the case if you made your frame from depleted uranium (which they pay you to take away), or a safer option (the danger coming not from the depleted radioactivity but from the environmentalists) – from material decoratively embossed with *cocacolacocacolacocac* which they pay you to promote. Then, at the same net cost of £8 (21.333 -13.333) you could convert your bedroom into a barber shop with a 2.666m by 2.666m mirror.

Chapter 8

Shapes

Close inspection of the title reveals that this chapter is called "SHAPES" rather than "Geometry". The reason for this is that shapes is what the chapter is about.

Most school material relates to shapes on planes, or "planar" shapes. "Geometry" means the study of measures of "Geo", or "the Earth", nowadays commonly believed to be round, yet the usual name for the topics we deal with here is still "planar geometry". (Flat Earth?...)

We will deal with shapes for three reasons: the third is that it is something that one is supposed to know about; the second is that it is good practise in 'multi-staged thinking', which is more important than knowing about 'geometry'; the first is that we need some results for the next chapter, which really is of practical importance.

8.1 Moving and turning: all you need to get into shape

Regarding shapes made of straight lines, there are only two things that matter: the relative length of the lines, and the angles between them. We must therefore be clear about what, exactly, these mean.

Where we are is called LOCATION. Changing location is done by MOVEMENT, which means covering a DISTANCE that is measured, for instance, in METRES.

The 'way we face' is called DIRECTION. Changing direction is done by TURNING. In doing so we go through an ANGLE, which is measured in DEGREES.

We can summarise this in the following table:

What we change	How we do it	What we go through	What we measure it by
LOCATION	MOVE ('along')	DISTANCE	METRES (etc.)
DIRECTION	TURN (rotate)	ANGLE	DEGREES

Just as we can face different directions from one location, so we can face the same direction from different locations, (specifically, also when the location is changed by moving not in the direction we face.)

Note: facing the same direction from such different locations does not mean that we are looking at the same point.

If you draw a straight line along the direction you face, and then move sideways and draw a second line in the same direction as the first, the two resulting lines are said to be parallel. Parallel lines thus have the same direction but no common locations.

In what follows, the word "turn" is used for changing direction only, in other words changing the way we face without changing location (as in 'spin' or 'rotate', not as in 'making a turn' while walking or driving).

We need to determine the size of 'One degree', that is, 'one unit of turning'. A 'complete turn' means turning until you face the same way as when you started (and 'half a turn' means turning until you face the opposite way). A complete turn is a lot of turning, so it is convenient to split the complete turn into a number measurable 'turning steps', the degrees. Dividing the turn into 100 would have been intuitive, but 360 was chosen instead. This, fortuitously, does have a reason: 100 can only be divided (without fractions) into 2, 4, 5, 10, 20, 25, 50, whereas 360 can be divided into 2, 3, 4, 5, 6, 8, 10, 12, 15, 18, 20, 24, 30, 36, 40, 45, 60, 72, 90, 120, 180. Good enough reason?

So one 360th of a complete turn is the accepted 'step' in turning. Instead of writing 'degree' or an abbreviation of this word after the number, (as in the unit of length 'cm') we use °. Thus half a turn (360/2) is 180°, and a quarter turn, as in from facing East to North, is 90°.

8.2 The eternal triangle

From here on we will deal mainly with the simplest closed shapes that can be made of straight lines: triangles. Studying the properties of triangles is important because even shapes with more than three sides are often best analysed by breaking them up into triangles.

As we have said, the main purpose of what we do here is to establish some facts which are required for the next chapter which deals with practical matters. On the way we will have an opportunity to practise multi-staged thinking, that is, exercises in which the final result is obtained through a sequence of mainly deductions, each serving the next one. Maths is useful training for this process which is so often required in other situations in real life. The individual steps themselves are not difficult, but many people are not used to persevering if the sequence has more than a couple of stages. We will proceed as though on a long walk: climb easy slopes with many steps. So, save your breath and your thinking will get fitter.

First we will use this process to obtain a striking result. Using what we understand about angles and parallel lines, we will show that the sum of the angles between the sides of a triangle is always 180° irrespective of the shape of the triangle.

8.3 Enter Art

Before we start an important point must be made, one that if omitted often makes students confused and lose heart. In most of what we have encountered so far in maths the required steps can be readily deduced by careful logical consideration. In the field of SHAPES (geometry), however, this is not always so, especially in the case of the first step. What is often needed is an idea. Ideas do not result from deductions, they are inspired. Here they often are about certain lines, which added to a drawing, provide the key to the solution. With practice these ideas come more readily, although their main source is of course imagination. This is where Maths touches on Art in all its beauty.

Note that at every stage we make use of and depend on results derived or proven in earlier sections. Also, attempting to follow the text without **constantly referring to the accompanying diagrams** is like watching Hungarian movies without reading the subtitles.

8.4 What a beautiful proof

Our first diagram is of a triangle - any triangle. The corners are referred to as A, B and O, and the internal angles of these corners as a, b and c. We will determine the sum of the internal angles.

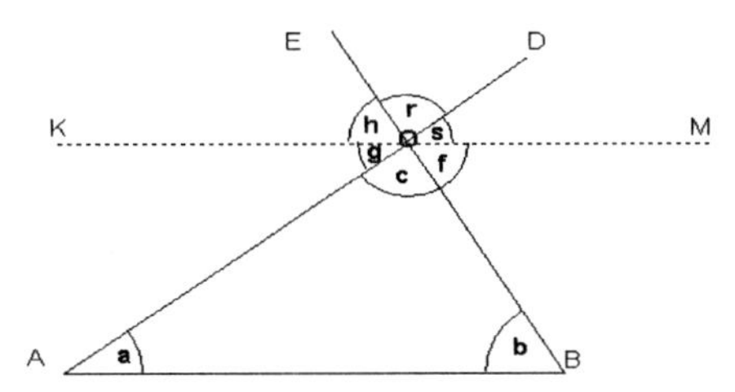

The idea (or 'trick', if you like) is to draw another 'auxiliary' line K-M through O and parallel to A-B, and to extend line A-O to D, and line B-O to E (refer every detail to the diagram and the marked angles).

Angle 's' is the 'amount of turning' from direction A-D to direction K-M. Angle 'a' is the amount of turning from directions A-D to A-B. We drew K-M parallel to A-B, i.e. running in the same direction, so turning from A-D to K-M is the same as turning to A-B, hence angles a and s are equal.

Similarly, turning from B-E to M-K through angle h is the same as turning from B-E to B-A through angle b, so b = h.

Finally, imagine a rigid dog standing at O, its nose pointing to D (and its tail to A). Now imagine the dog turning until its nose points at E, in other words its nose rotated through angle r, and its tail points to B, having rotated through angle c. Being a rigid dog, its tail-end turns as much as the nose end, so it follows that angles c and r are equal.

The sum of the internal angles of the triangle is $a + c + b$.

As $a = s$, $c = r$ and $b = h$, the sum $a + c + b$ is equal to $s + r + h$, which is the angle covered by turning from O-M to O-K, which is half a complete turn, or 180°!

Thus we prove that $a + b + c$, which is the sum of the angles of the triangle, of any triangle, is 180°

While we are at it, we can show another useful result. If we copy just part of the diagram above, with K-M still parallel to A-B

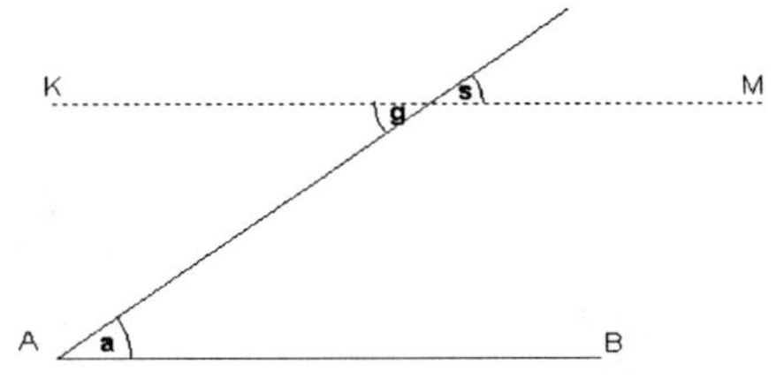

The 'rigid dog' will confirm that $g = s$; we have already established that $s = a$.
So, we can deduce that $g = a$. (and in the same way that $b = f$ in the first diagram).

Angles g & a (and b & f) are called alternate angles. (Why not alturnate?...)

A triangle is defined by the lengths of its sides and the size of the angles between the sides. With a given triangle, unlike an aquarium, it does not matter which way up it is: turned over, rotated, it is still regarded as the same triangle.

8.5 Fitting tests

Need all three sides and angles be specified to define a triangle?

Prepare a few narrow cardboard strips of different lengths; take four of them and join their tips with upturned drawing pins to form a four-sided shape.

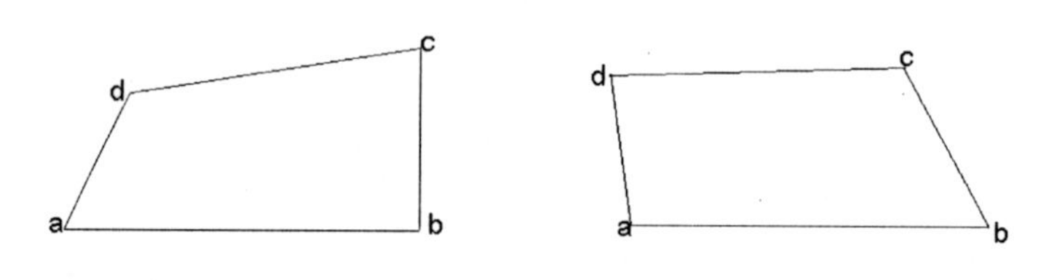

Holding down one of the sides (a-b), the two free corners (c & d) can be moved about, so you can change the shape.

The same goes for shapes of more than four sides, but make a three-sided shape (triangle) this way, hold down one of the sides, and nothing else can move! (That is why strong structures such as bridges are made of triangular sections.)

This means that if the lengths of the three sides of a triangle are fixed, nothing further needs to be specified about the angles, only a single triangular shape can be made. Any number of triangles made of three sides of the same lengths as those of the original model will all be found, maybe only after rotating or flipping them, to fit exactly when laid one on top of each other (meaning that all sides and angles in one triangle are same as in the others). Such triangles could be called 'equal' triangles. If, however, this makes it too clear, you can always use the conventional term - 'congruent' triangles.

There are other ways in which to define the shape of a particular triangle (that is, to establish the relative positions of the three corners).

Let us examine what happens if we set only two of the lengths of a triangle: lengths L_1 and L_2, and the angle v between them.

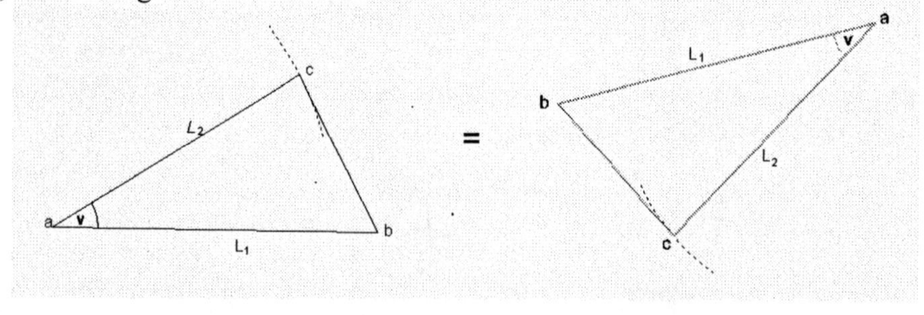

The extremities of L_1 define two points (a and b). If one end of L_2 is at 'a' we know that the other end, point c, must be somewhere removed from 'a' by distance L_2."Somewhere L_2 away from a" means some point on a circle the centre of which is 'a' and the radius is L_2.

If we now fix the direction of L_2 relative to L_1, in other words fix the angle (v) between them, we determine the position of the third point (c) on that circle. Thus we have the relative positions of the three points and the triangle loses any freedom (to 'distort').

There is yet another way in which to define a triangle (that is, to fix the relative positions of its three points). By setting one length, L, the ends a b of which fix two points, and the angles v and u, which determine the direction of the two lines extending from a and b. There is then only one point at which these two lines can meet, and it defines the third point c.

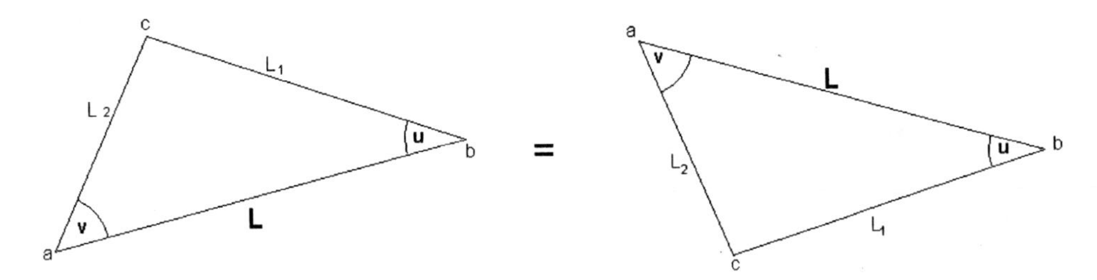

So, the relative positions of the three corners of the triangle are fixed by the length of one line and the two angles which direct the other lines out of the ends of first one.

To summarize: Two triangles are equal (i.e. all three sides and three angles in one triangle being equal to those in the other triangle) if any of the following sets of three items only, are known to be equal:

3 sides	2 sides and 1 angle (set between the two sides)	2 angles and 1 side (set between the 2 angles)

What of 3 angles?

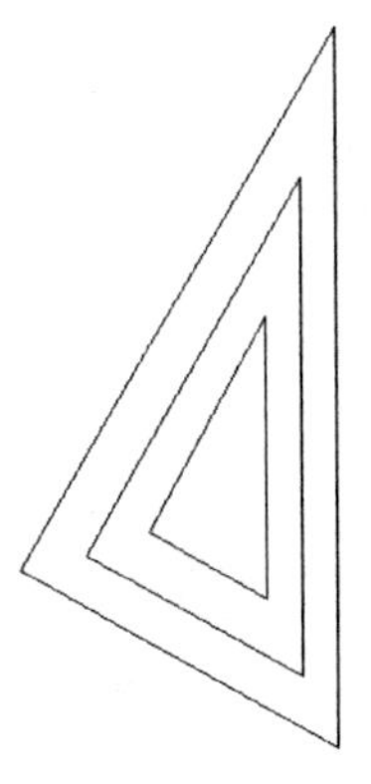

One can increase ('inflate') a given triangle by 'shifting' its sides outwards without changing their directions. If the directions of the lines do not change, nor do the angles between them. Therefore, when two triangles have the same three angles the triangles are not necessarily equal, because their sizes (i.e. the lengths of their sides) may be of different.

Such triangles, with all angles in common, are called ‘similar’. (Thus equal triangles are always similar, but similar triangles are not all equal.)

Note: when triangles have two angles in common so too is third one, (because the sum of the three angles is always 180).

When several similar or equal triangles lie in different orientations one must note carefully which angles and sides in one triangle correspond to those of the other.

Similar triangles:

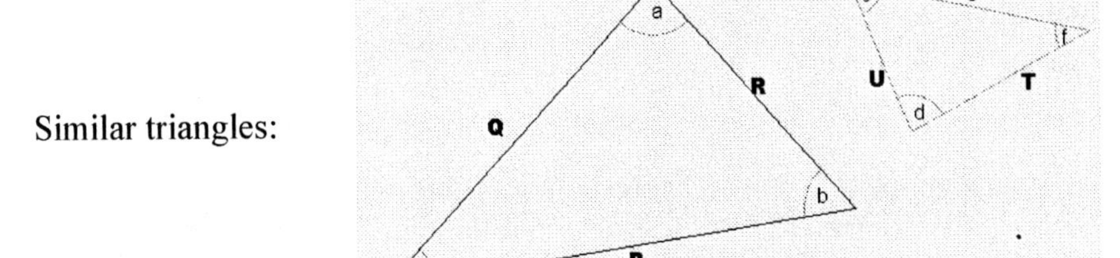

Here the equal angles are: a = d; b = e; c = f.

Two sides (one from each triangle) that lie opposite the equal angle are called the ‘matching sides'. Here the matching sides are: P (opposite a) & S (opposite d) (a=d); Q & T; R & U.

In equal triangles it is the matching sides that are equal.

A 90° angle is called a 'right angle'; (this is not in apposition to 'left angle' or 'wrong angle'...). The concept of two lines with a 90° angle between them is blessed with a particularly generous allotment of confusing terminology: 'perpendicular', 'normal', 'orthogonal' etc.; we will just say " at 90° ".

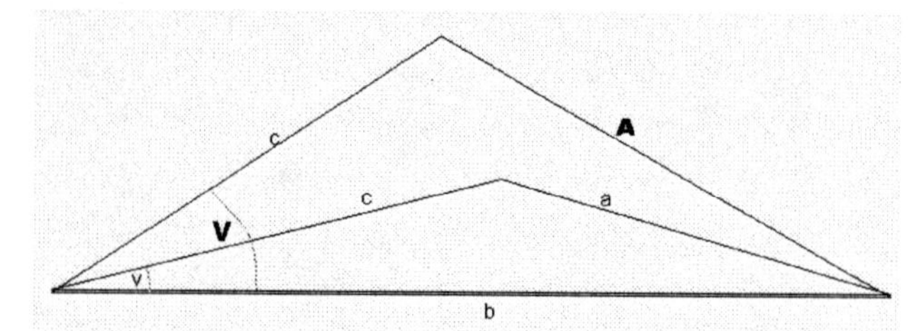

A is larger than *a*, so angle *V* is greater than *v*

For adjacent sides of given length (b and c) the size of angle v between them obviously depends on the length of side a opposite that angle. If all the sides of a triangle are equal, all three angles sit between sides of equal length, and also opposite to sides that are equal. There is thus no reason for any of the angles to be larger than any other, so all three angles must be equal.

We know that the sum of the angles of a triangle is 180°, so when the sides of a triangle are all equal, each of its (equal) angles must be 60°. We will call this triangle a '3-equisided' one.

Let us now consider triangles in which two sides are equal. We will refer to them as '2-equisided' because the alternative would be the conventional name 'isosceles'.

(We can use the abbreviations Δ for triangle, ∡ for angle, and ‖ for "parallel to").

We will show that in a triangle of this kind, for instance ABC with sides AC = BC, the angles a and b are equal. How?

We could, of course, simply say that

the situation of ∡a between AB and AC, and opposite BC

mirrors

the situation of ∡b between AB and BC, and opposite AC.

However we will prove this using a more general method, one that is commonly used when the equality between certain sides and/or angles is given, and the equality of other sides/angles needs to be proven.

We need to find a link between the equalities that are given and the items that we want to prove to be equal.

8. 6 General tactics for proving equality

Let us first look at an example of such an exercise in a different area. Let us suppose that we have two car-wheel bearings; we want to prove that they are identical without having to take them to bits. We could do so by establishing that both bearings came from identical cars. To establish the equality of the cars, we would not need to compare all their details; it might be enough to learn for instance that both cars look like slippers and have divine seats: can only be the old Citroen DS, so we know that the other details of the cars are the same, including their wheel bearings. In this case the cars provided the link between the given equalities (the 'beauty' & comfort) and the items (the bearings) whose equality we want to prove.

In the same way, a partial description of two triangles can establish their equality; we know that of the six elements of a triangles (three sides and three angles) only three need be shown to be equal in the two triangles for their other elements to be proven equal too.

If we have a pair of sides or angles whose equality we want to prove, and we also have certain other pairs of sides and/or angles given as equal, we can look for two triangles to link between all these: each triangle containing one member of the pair which we want to prove equal, and one member of each pair that is given as equal. Then, using these equalities that was given for some of the elements of the two triangles, we attempt to prove that the entire triangles are equal. If we succeed, we have also proved that the rest of the elements are equal, including those whose equality we wanted to prove.

If the diagram does not contain two such triangles, we can create two triangles that fit the above requirements by adding 'helping' lines (as we have already encountered). Each of the triangles created in this way must contain:

one of the pair of items we want to prove are equal

one of each pair of items given as equal

the new line or lines drawn in such a way that they add further equality between the triangles

As we mentioned before, drawing the line in such a way that it serves our purpose sometimes requires inspiration, or an idea, although it is usually quite a simple one.

Let us now apply this method to prove that $\measuredangle a_1 = \measuredangle a_2$ in the2-equisided triangle CAB (with CA=CB):

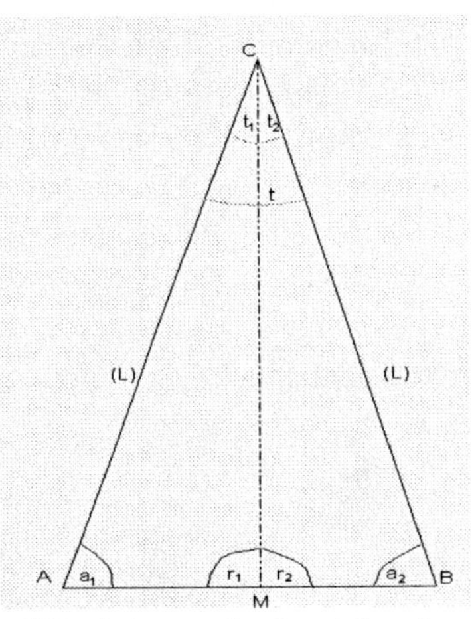

We said we need two triangles; in this case the 'idea' is to add a line from C to meet line AB at some point M, and to draw this line so that it divides angle t (between CA and CB) into two equal parts: t_1 & t_2. We now have the two triangles, AMC & BMC, each of which contains one of the items which we want to prove equal, namely $\measuredangle a_1$ & $\measuredangle a_2$, and one of the items which were given as equal: CA & CB (i.e. Δ CAB given as 2-equisided).

The way the added line was chosen contributed two further equalities between the two triangles:

- $\measuredangle t_1$ of Δ AMC which equals $\measuredangle t_2$ of Δ BMC, and
- (very simply) line CM which is common to the two triangles.

With two sides (CA=CB and the common CM) and one angle equal, we know that the two entire triangles are equal.

We know then that the rest of the angles are equal in each triangle: $\measuredangle a_1$ and $\measuredangle a_2$ each are opposite the same side CM, so they are 'matching angles' and thus are equal; this is what we set out to prove.

The remaining angles r_1 & r_2 are equal too. And since together they make 180° each must be 90°.

AM and BM are matching sides (they are opposite to equal $\measuredangle$'s t_1 & t_2), and so they must be equal. So, in the process we have also learned that in a triangle that has two equal sides a line that bisects (cuts in two equal parts) the angle between the equal sides meets the third side in the middle, and it meets it at 90°.

8.7 Three importantly special triangles

Now we proceed to discover something less obvious.

Consider Δ ABC which is now a 3-equisided triangle; now also AB equals AC & BC. All three angles- t, a_1 and a_2 (which we already found to be equal to a_1) are, as we know, also equal, at 60°.

All the equalities in the previous drawing of the 2-equisided triangle are of course unaffected by making also the third side (AB) equal to the other two:

$\measuredangle t_1$ is still half of $\measuredangle t$; As $\measuredangle t$ is now 60° $\measuredangle t_1$ is now 30°.
($\measuredangle r_1$ is still 90°)

Also, since AM is still half of AB it is now also half of AC because AC = AB).

So if we consider only half of triangle ABC, i.e. ΔAMC (now figured separately), we have a triangle with angles of 90°, 60° and 30°:

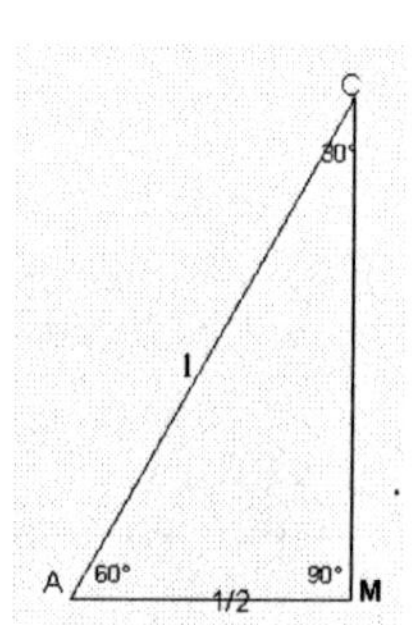

in a triangle of this type the side opposite the 30° angle (AM) is half the length of the side opposite the 90° angle (AC), or, half the longest side (longest, because it sits opposite the widest angle). In other words, in a triangle with a 90° and a 30° angle:

the ratio between the side opposite the 30° angle and the longest side is 1/2.

Now let us look at a triangle in which the angle between the equal sides is 90°;

(it makes no difference that here we rotate the triangle

from

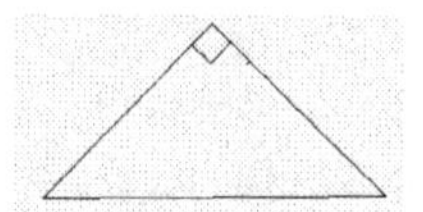

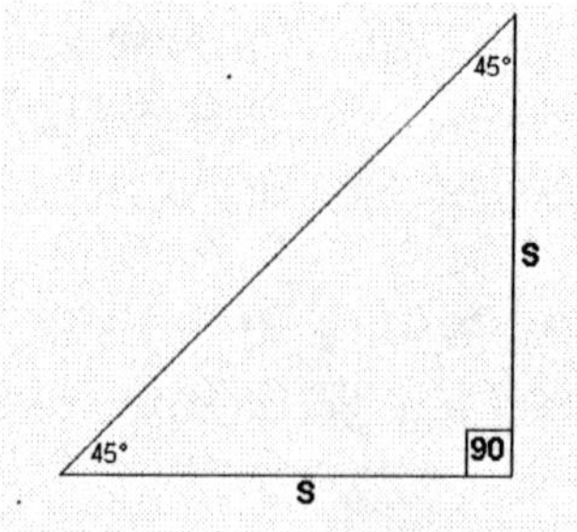

to 'lie' it on one of the equal sides)

Again, the other two angles are equal. Together with the 90° these two angles add up to 180°, so between them they must add up to 90° which means that each one is 45°. In other words, in a triangle with angles of 90° and 45°, the two 'at 90°' sides are equal, or, the ratio between them is 1.

8.8 And a special 'fourangle'

We will next look at an important 4-sided shape in which both pairs of opposite sides are parallel.

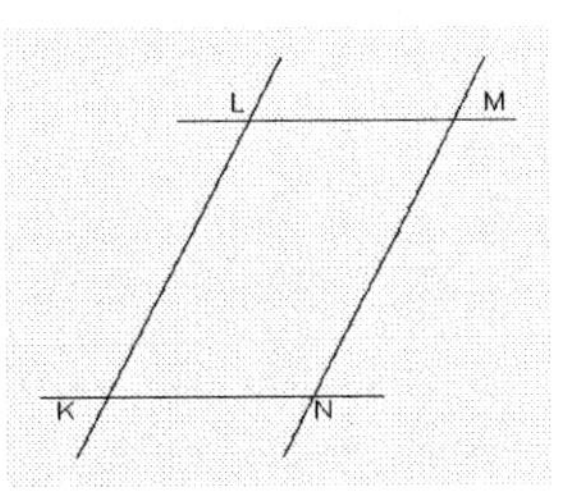

KL is parallel to NM and KN is parallel to LM. This shape, surprisingly, has a sensible name: parallelogram. It has an interesting property; the lengths of the opposite sides are always equal.

To prove this we resort again to triangles, by adding a 'helping' line to the figure. In this case little inspiration is needed; it is quite easy to deduce where a promising line can be drawn.

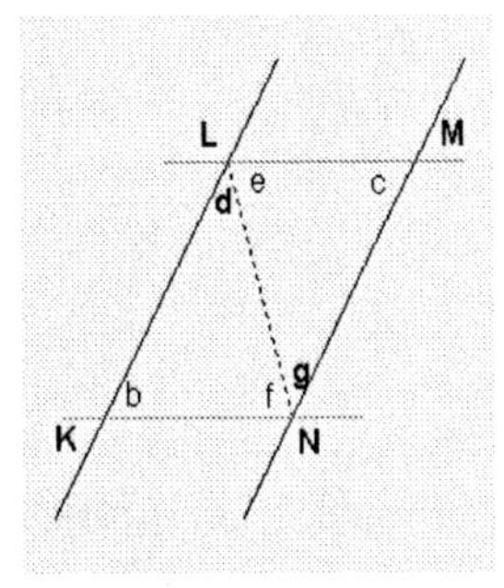

We want to prove that

LM = KN

and that MN = LK.

This result is most readily obtained if we can create one triangle containing LM & MN and another triangle containing KN & LK, and then show that the two triangles are equal. The line that achieves this is obviously LN, creating triangles LNM and NLK.

LM & KN will of course have to be the matching sides, and so with MN & LK. (The reason for this precaution is that in theory the two triangles could also be equal with LM in one triangle being equal to LK in the other, rather than to KN, as we want to prove.)

As elsewhere in this book, the purpose now is not to teach how to prove that these two triangles are equal, but to show how to deduce the method for doing so.

We know two relevant things:

1. We can achieve nothing unless we make use of the given information. In this case it is the parallelism of opposite sides.

2. As a general rule we need to show that one of the following sets of items is common to the two triangles: 3 sides, 2 sides and 1 angle, 1 side and 2 angles. In this case there are two sides in each triangle (LM & MN in one triangle and KN & LK in the other) which cannot be used to prove the equality of the triangles: the whole purpose of proving the equality of the triangles is, after all, to show that these two pairs of contained sides are equal.

With only one side left with which to prove the equality of the triangles, we conclude that we can only use the set of 1 side and 2 angles.

As to the remaining side – the one we need to show that it is equal in both triangles, there is not much to ponder about: it is of course LN which is common to both triangles.

All that remains to do is to find two pairs of angles that are equal between the two triangles; we would do well to use the fact that LM is parallel to KN, and KL to NM.

(for greater clarity, we copy only the

relevant parts of the drawing:)

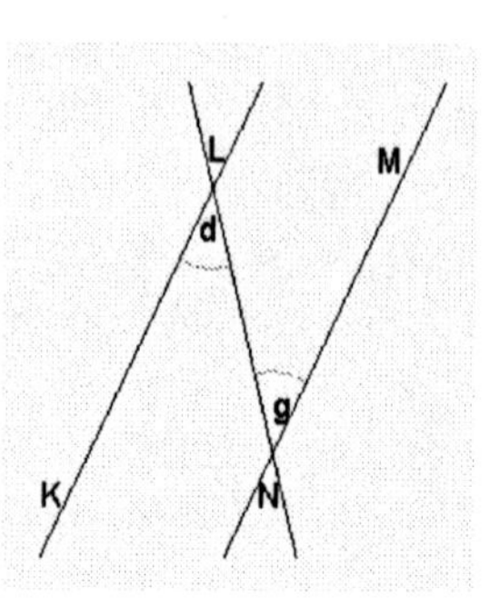

Looking at parallel lines KL and NM and the line LN which crosses them we recognise angles d and g as the alternate angles and therefore know that they are equal. Similarly, looking at parallel lines LM and KN below, again crossed by LN, angles f and e too are now (equal) alternate angles.

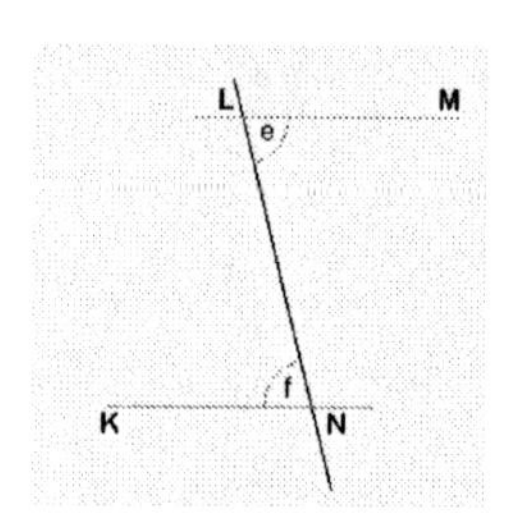

So, (referring back to the full diagram on the previous page) in triangles NLK and LNM we have:

$\measuredangle d = \measuredangle g$,

$\measuredangle f = \measuredangle e$,

and their common side LN.

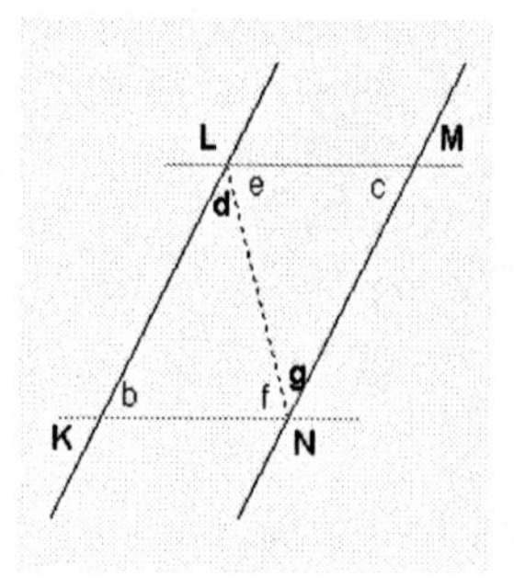

Therefore triangles NLK and LNM are equal.

All that is left is to make sure what is equal to what in here, in other words which are the matching sides. We remember that two sides in different triangles match if they are opposite equal angles.

$\measuredangle f$ of $\Delta NLK = \measuredangle e$ of ΔLNM,

so, as side LK is opposite $\measuredangle f$, and NM is opposite $\measuredangle e$, these two sides are matching, thus equal.

$\measuredangle d$ of $\Delta NLK = \measuredangle g$ of ΔLNM,

KN is opposite $\measuredangle d$, and LM is opposite $\measuredangle g$, so also KN and LM are matching, thus equal.

Looking back at parallelogram KLMN, we see that

LK is the side that is opposite MN and we proved that it is equal to it,

KN is the side that is opposite LM and we proved that it is equal to it.

This, you might remember, is what we set out to prove.

8.9 The very important feature of similar triangles

We are now ready to prove a result which is crucial to the important material in the next chapter:

In similar triangles the ratio between any pair of sides in one triangle equals the ratio between the matching pair of sides in the other triangle.

The proof is somewhat like jogging: healthy exercise and not difficult, just a little low on excitement, but even if you live for pleasure alone it is worth going through. It does not take long, and it is a good example of how important results are arrived at through a chain of simple logical steps. The proof is quite simple, but deducing how to come up with the method is not that obvious, so, exceptionally, we will commit the pedagogic crime of merely showing how it is done.

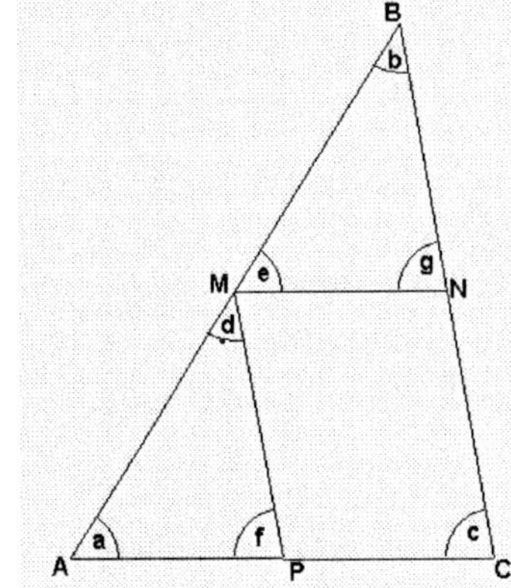

Let us consider triangle ABC. Half way up AB we mark a point M and draw a line MN parallel to AC, and another line MP parallel to BC.

(The following description requires close attention and constant reference to the diagram).

Looking at angles a, b, c of ΔABC and angles a, d, f of ΔAMP we see that ∡a is common to both triangles; ∡d = ∡b and ∡f = ∡c, both because MP ‖ BC.

So, with all three angles the same between both triangles, **ΔABC and ΔAMP are similar**.

We have just noted that ∡b = ∡d. Also ∡e = ∡a because MN ‖ AC. Sides AM = MB because we put M in the middle of line AB.

So, considering ΔMBN and ΔAMP we find that they are equal because they have two angles and one side equal. We needed this result for showing that MN = AP.

Now, since MN ‖ PC and MP ‖ NC, MNCP is a parallelogram, and thus MN = PC, i.e. AP = PC.

As AC is the sum of the two equal portions AP and PC it follows that AC = 2AP ... (1)

Also, from the parallelogram: MP = NC, and from ΔAMP = ΔMBN it follows that: MP = BN.

As BC is the sum of BN and NC, each of which is equal to MP, it follows that BC = 2MP ... (2)

Finally, as M was marked in the middle of AB, we know that AB = 2AM ... (3)

Note: on the left of the three equations are the sides of ΔABC and on the right those of ΔAMP.

So far, then, we have proved that if there is ratio of 2 between any two matching sides (one from each of two similar triangle), then this ratio holds between all the matching sides.

From this we can now deduce the related result which we are after, namely, that the ratio between any two sides within one triangle is the same as the ratio between the (matching) pair of sides within the other, similar, triangle

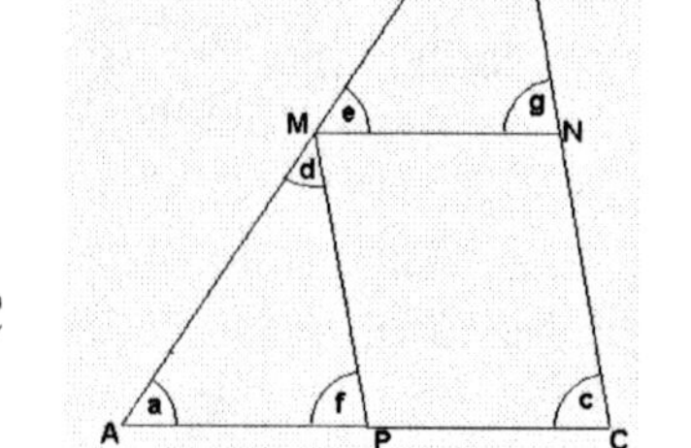

i.e. that $\dfrac{AC}{BC} = \dfrac{AP}{MP}$, $\dfrac{AC}{AB} = \dfrac{AP}{AM}$, $\dfrac{BC}{AB} = \dfrac{MP}{AM}$

(the drawing still yearns to be looked at).

(Note: the left side of these equations deals with sides of ΔABC and the right with those of ΔAMP).

Two sides of an equation remain equal if both are divided by equal amounts.

Therefore, if we take AC = 2AP (eq.1), the equality will be upheld if we divide the AC by BC and the 2AP by 2MP, because eq.2 states that these two dividers are equal,

i.e. $$\frac{AC}{BC} = \frac{2AP}{2MP}.$$

Multiplying by 2 and then dividing by 2 means we needn't bother about either, and so we are left with the first of the above results.

We get the second of the above results if we follow the same procedure using eq.1 & eq.3 (AB = 2AM); and the third one by using eq.2 & 3.

We did prove what we wanted, but admittedly only for similar triangles where one was twice the size of the other. For similar triangles with other relative sizes, the process is the same, except that AB is divided into a greater number of (equal) sections:

To summarise:

In the triangles

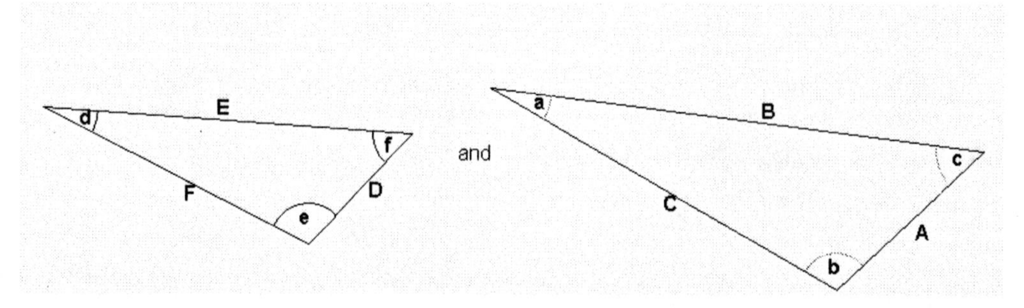

if ∡a = ∡d, ∡b = ∡e (and ∡c = ∡f), then C/A = F/D

C/B = F/E

B/A = E/D.

Chapter 9

Independent Components

You might recall from the darkest days of your distant past that in a triangle with sides of lengths A, B and C, with an angle of 90° between sides A and B, and α being the angle between B and C:

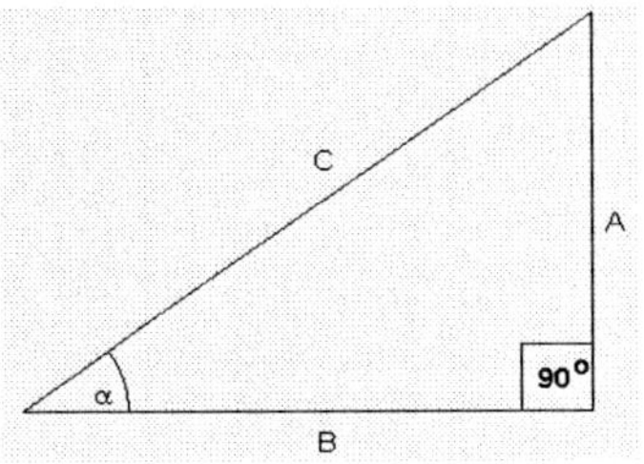

the ratio between A and C, i.e. A/C, is called the “sine” (of) α

the ratio between B and C, i.e. B/C, is called “cosine” α

and A /B is “tangent” α, and

that the values of these ratios were tabulated for all angles α in what was called “trigonometric tables”, and nowadays pop up when one presses correspondingly marked keys on calculator.

This raises three important questions that are seldom explained and even more seldom asked.

Why are the tables only for triangles with an angle 90° (between A and B)? Why not for triangles in which this angle is 100°, also a nice number… What is so special about 90°?

Why tables only for *ratios* between the sides; why are there no tables (and names) for their sums, or for their ‘products’ (i.e. one *multiplied* by the other); what is so special about the ratios?

Why do these ratios have such silly names, like secret codes designed to confuse an enemy?

We have no answer to the third question - only commiseration, and a suggestion for better alternatives to these terms. There *are* however very good answers to the first two questions. In fact, if one does not address them there is little point in spending so much time and attention on this topic altogether. A properly educated student should refuse to enter into all this without knowing what its purpose is.

9.1 Holy 90° - Why?

Let us begin with the '*90°*'. There is indeed something particularly important about this angle, because it reflects a very basic logical notion. Consider the following five pairs of concepts:

1. Atoms and molecules in a solid object are ‘bound’ to each other. This does not mean that they are immobile; on the contrary, they vibrate continuously over their mean position; they may vibrate lightly or vigorously.

The object may be hot or cold. The more vigorously the atoms vibrate the hotter the object is. It is not that heat is *related* to the strength of the vibrations: heat *is* the vibrations. There is no such thing as 'heatness', which dwells inside the object and grows when the atoms vibrate harder. When you touch an object that feels hot all that happens is that the object’s atoms are hitting the nerve ends in your finger.

Here then we have two concepts: the degree of 'restlessness' of the atoms/molecules of an object, and how "hot" the object is. The two are essentially the same thing, at least in the sense that one can never change without the other changing too.

2. Terms such as "Beauty" and "Allure" (or "Age" and "Old") are concepts that are closely related and more or less dependant, though they are *not entirely* so.

3. "Colour" and "Odour" (or "Colour" and "Height") are entirely different, 'independent' concepts. A change in one does not in itself affect the other.

4. "Pitch" (of a musical note) and "Loudness" are generally independent, but not *completely*: high notes tend to sound a little louder to our ears.

5. "Hot" and "Cold" are not independent at all; they are very much of the same kind, but they act as *opposites* of each other.

Now let us look at some pairs of entities which have an angle between them.

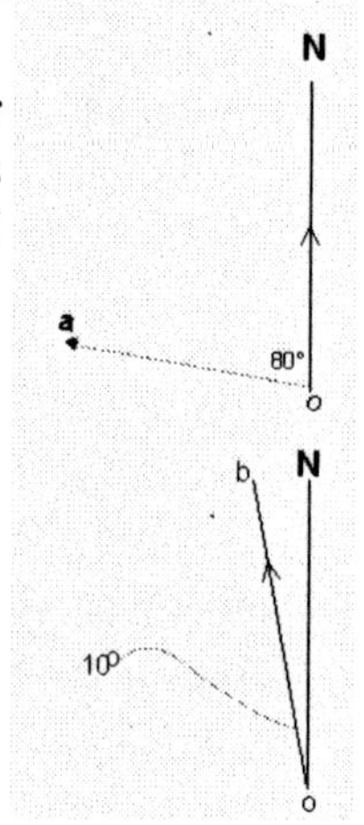

Consider going from *o* due North, the other going from *o* to *a*. The latter does not do *much* by way of taking us North, but it does so a little. This is comparable to the situation described in (4) above. The angle between N and *o* is a little less than 90°, say, 80°.

Now we have a line leading from o to *b*; here our progress is mainly northward, but still not entirely so. The situation is comparable to that described in (2) above. The angle between N and *o* here is small, e.g. 10°.

However, by travelling due West (*o* to W) we are doing *absolutely nothing* to increase our chances of meeting a polar bear. We get no nearer, and no further, from North. This is akin to the situation in (3), involving *independent* entities, or *independent 'factors'*. The angle between *o* - N and *o* - W is 90°!

p - *m* is parallel to *o*-N, which means that the angle between it and *o*-N is 0°. When we go from *p* to *m* we are travelling due North, as from *o* to N, nothing else. This is comparable to the situation in (1).

Finally, if we go from *o* to S we invest our entire effort in travelling *away* from North, or making *negative* northward progress, which is the same as doing the *opposite* of going North. This can be compared to case (5). The angle between *o*-N and *o*-S is 180°.

What this tells us is that lines that are at 90° to each other are a graphic representation of *mutually independent* factors, that is, independent *of each other*. The reason this topic is of such importance is that the solution of complex problems in *any* area must start with analysing the problem into its mutually *independent* factors.

Failure to do this leads to going around in circles. A typical problem arises when, carelessly, several pertinent factors are named separately while they are, at least in part, 'the same thing called by a different name', in other words: factors that are not *independent* of each other. Adjusting one of the factors, therefore, alters others, and vice-versa, hence the 'going in circles'.

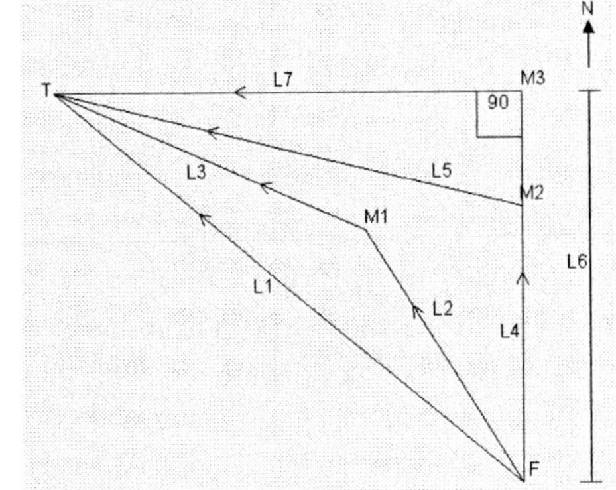

Let us take an example:

We want to find the cost of going from F to T.

There are several ways we can get from F to T: we can go direct; we can go from F to M_1 and from M_1 to T, or we can go via M_2. All these routes amount to the same thing because they get us from F to T.

But let us suppose that any movement northward costs us £C per Km (such that the total cost of a northward section of a length L would be $L \times C$; also, that for any *southward* movement we are given a *refund* of £C per Km). East-west travel is free. Note already here, that if we want to make some travel free it *must* be at 90° to the northward one, because otherwise it would contain *some* northward movement and thus could not be free.

The direct route, length L_1, does not allow us to calculate the cost because it does not lie entirely in the northward direction and so not all of L_1 is chargeable at the given rate. To go via M_1 is no better, because, like L_1, neither sections L_2 and L_3 go entirely northwards. Going via M_2 consists of a purely northward section L_4 the cost of which we *can* calculate ($L_4 \times C$), but this still leaves us with L_5 which, like L_1, L_2 and L_3, involves an unclear mixture of chargeable and free travel.

If we go via M_3 however, section L_6, from F to M_3 lies solely in a northward direction, and section M_3 to T solely in the free westward direction, with no northward or southward movement. This route is thus divided into two sections that *are* useful for calculating cost:

L_6 leads purely northward, and its cost can thus be calculated: $L_6 \times C$.

L_7 involves no expenditure, because in includes no northward or southward movement.

Therefore the cost of L_6 contains the *entire* cost of the journey ($L_6 \times C$).

What makes this separation useful is that one of the sections, L_6, is *entirely* in the direction of interest (the chargeable one), and the other, L_7, being *at 90°* to the first one, has nothing to do with the special interest and can therefore be ignored. On the other hand, L_2, L_3 etc. are useless mixtures of the relevant and 'ignorable' components.

This means that if we represent an activity by a directional line A→B, and then replace this representation with two other, consecutive lines A→C and C→B which are at 90° to each other (i.e. ABC forming a triangle containing a 90° angle at C), we obtain two directional lines which represent two *independent* parts, or *components*, of the activity.

This is often done, as in the example just given, with the purpose of 'extracting', or isolating the one component that contains all (and nothing but) the object of our interest.

Let us take another example: a wheeled cart on a sloping road.

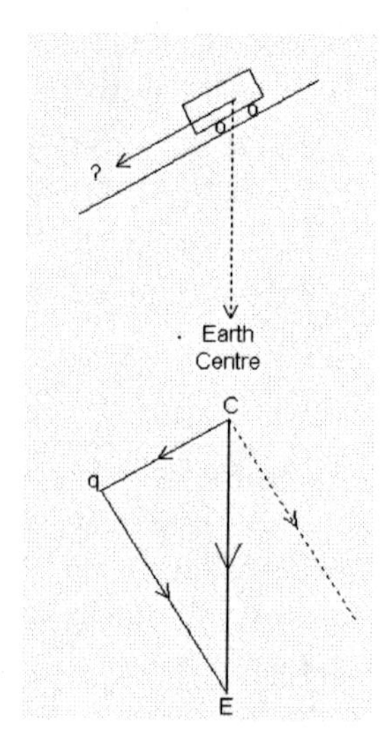

The cart is subject to a force: Mother Earth pulling it vertically downward to her centre (gravity). Roads are hard, so gravity cannot pull the cart underground. The only way the cart can move is down the sloping road. We want to know what proportion of the earth's pull is available for drawing the cart in the only direction in which it can move.

We can tackle the problem in the same way as in the last example in which the move from F to T was achieved by replacing it by two other moves. First we represent the 'source' force by a line C-E in the (downward) direction of the force, and of a length that represents the amount of force (on some scale, for instance, 1mm per Kg).

As in the last example, the 'source' force C-E can be completely represented by two other, distinct forces, for instance by a fellow pulling in direction C-q, the force being represented the length of the line C-q, and another fellow pulling in direction q-E, the force again represented by the length of the line.

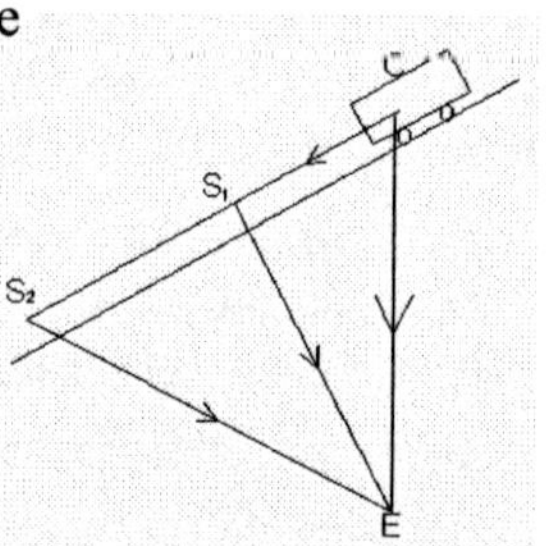

(In actual fact they would both have to do their pulling at the same point, C, for reasons that we will not go into here; what matters is that the *directions* and *strengths* of the forces are still represented by the lines of triangle CqE.)

Back to our cart. Gravity is represented by a vertical line C-E. What interests us is the force along the slope. We represent this force by line C-S, but how long should we draw the line?

We have found that we can replace the earth's pull C-E with a pair of forces, also represented by lines which, together with line C-E make up a triangle. Here, naturally, one of the pair of lines is C-S, in the direction of our slope, and the other runs from s to E. But C-S might extend to any of many points: S_1, S_2 etc.. Let us look at C- S_2-E as we have drawn it: this is no good, because while C- S_2 does show a pull down the slope, it is obvious that S_2-E contributes to some extent to pushing the cart back up the slope.

What we need is for line C-S alone to represent the *whole* force that drags the cart along the road (and nothing else), and for line S-E to have no part in it (neither forwards nor backwards). But how do we achieve this?

By drawing C-S and S-E at 90° to each other!

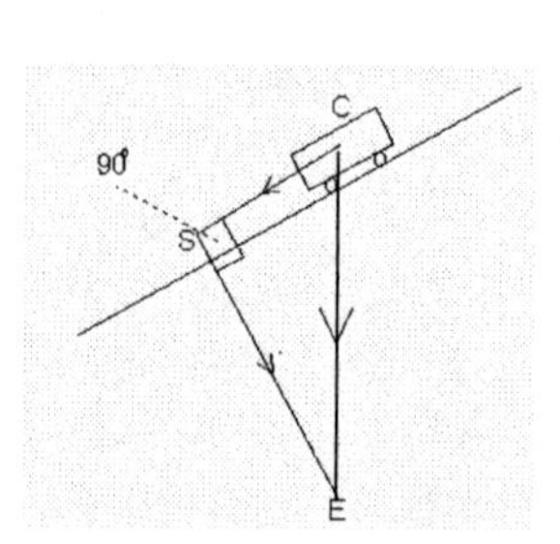

Here again we have split an activity (C-E) into two '*components*', and to make sure that the components are *mutually independent*, in other words that one component takes no part in what the other is doing, we put them at 90° to each other.

That is what is so important about 90°.

(In the following "90° triangle" will mean "triangle with one of its angles 90° ")

9.2 How large are the mutually independent components?

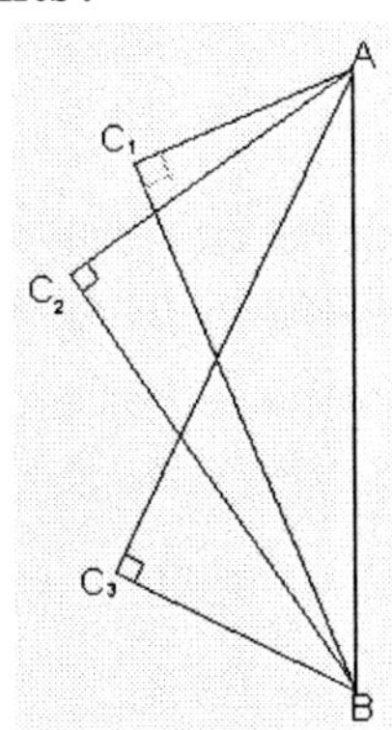

There are many ways in which we can split an activity (which we represent by a directional line) into mutually independent components, as shown in the diagrams here. For a given 'activity line' A-B many triangles can be drawn with an angle of 90° (∟) between the other two ('component') lines, for instance triangles A-c_1-B, A-c_2-B etc..

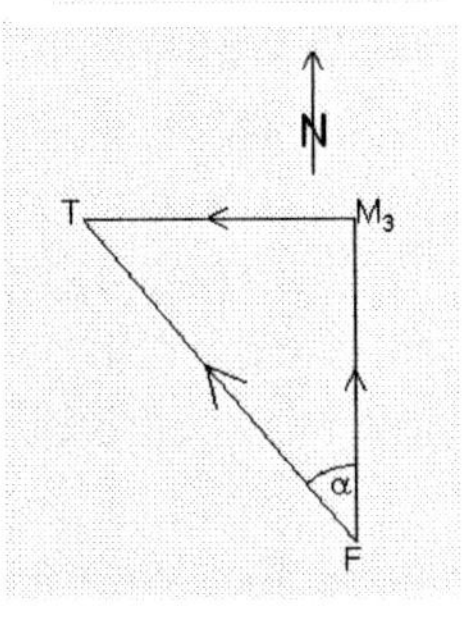

Which of these we want depends on what direction we need one of the components to be in. This is established according to the context of the case. In our first case above, in which we wanted to know the length of the northward component of a given movement F-T, the required component (F-M_3) simply pointed Northwards. In the case of the cart the direction of the required component (C-S) was down the road.

We will use α to denote the angle between the component of interest and the 'source activity', F-T in the first example, C-E for the second.

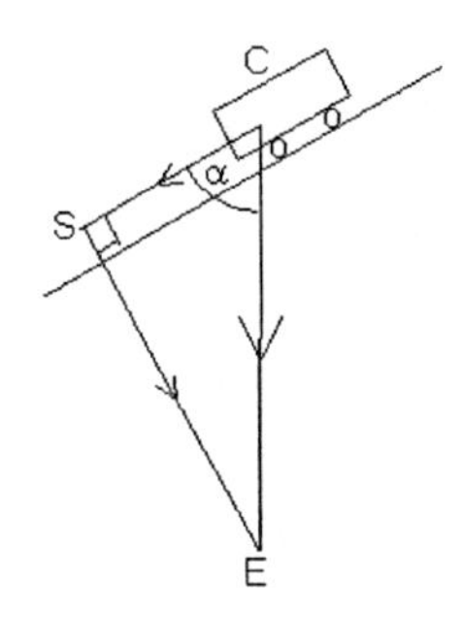

Looking again at a diagram of various 90° angles A-c_1-B, A-c_2-B, A-c_3-B 'resting' on a given 'source' line A-B, we see that the length (and hence the strength) of the required component A-c depends on the angle α between it and the 'source' A-B. Specifically, we see that the smaller this angle the larger the component along the angle. This makes sense: if the direction of the component is fairly close to that of the source (where angle α between them is narrow), we would expect the component (A-c_3) to be not much smaller than the source.

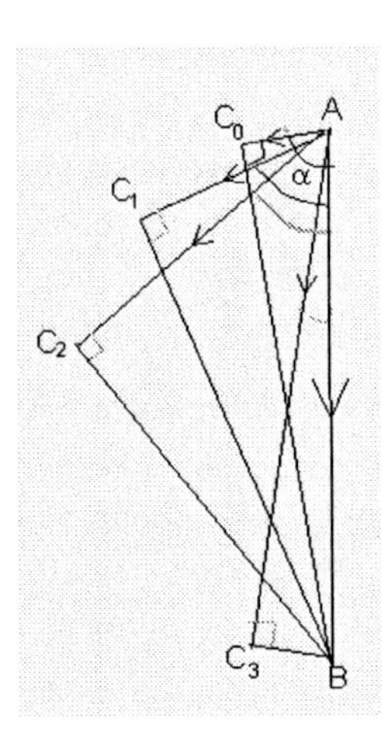

A component can of course never be *larger* than its source, (hence the source is always the longest line in these triangles with 90° between the component lines). When ∡α becomes nearly 90°, i.e. the component nears a direction which has 'nothing to do with the source', the component (A-c_0) tends, as expected, to zero length.

Therefore if we know the angle between a source of a given length and a component we should be able to establish the size of the component. But how?

Our task is to find the size of component C, given the size of source S and the angle α between the two.

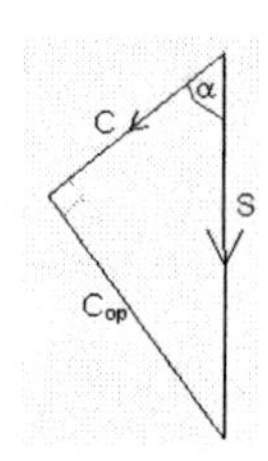

If we know S, and want to know C, it would of course be helpful to know the ratio between these two i.e. C/S (we will call this ratio 'R' for short).

There are two questions: where do we get the value for this ratio, and how exactly do we use this ratio, namely: what do we do with the given S and the ratio $\frac{C}{S}$ to produce the required C?

We will answer the second question first because it is short and simple. Naturally the way to get C from S and $\frac{C}{S}$ is to multiply the two together, i.e. $S \times \frac{C}{S} = C$.

If we use 'R' for this ratio C/S, then **C = S × R**

Now we need to establish from where we can get the value of this ratio R (or C/S).

We remember that with α given, *all* the angles of the triangle are fixed: apart from the given α, a second angle is known because S is always replaced by two *mutually independent* components (one of which is C), and the angle between them is 90°; the third angle has no freedom as it must be 180 less the other two, i.e. $180 - 90 - \alpha$ $(= 90 - \alpha)$.

We also know that in a triangle with given angles the *ratios* between the sides does not depend on the *size* of the triangle. (Yes, here is where we make the promised practical use of results that we prepared in the 'Shapes' chapter.)

Therefore if the angle between the source (S) and the component (C) is given, the *ratio* (R) between the two is fixed, irrespective of the size of the source.

This means that once R is found in any one triangle with 90° and α it can be used as the ratio between sides C and S in any other, larger or smaller, 90° triangle with α between these sides.

Given this, all we need now is some patient soul to sit down and, once and for all, draw triangles with every possible angle α, all also having a 90° angle, (he can draw the triangles any size he likes); then, in each triangle measure the longest line and the other line next to angle α, and divide the latter by the first, and so work out R for every angle α.

The phrase "Ratio R for angle α" is shortened to "$R(\alpha)$ ".

Fortunately this labour of love has already been accomplished (more than five hundred years ago). Having gone to all that trouble, he then wrote it down in the form of a 'table' containing a list of all α angles, and their $R(\alpha)$.

So now, whenever we have a source activity S, and want to know the size of its component C in a direction $\alpha°$ relative to S, we can look up the $R(\alpha)$ in the table (or squeeze them out of a calculator) and multiply S by it! For example, in the last diagram S happened to be 25mm, and angle α is 53°, for which R is (approximately) 0.6, therefore C ought to be $25 \times 0.6 = 15$ mm, which it is.

With all the triangles at hand our medieval benefactor took the opportunity to measure and calculate the ratios also between the other component, the one opposite to α (i.e. C_{op}) and S, and so compiled a table also for these ratios $R_{op}(\alpha)$ for each angle α.

So now, following the same reasoning as before, when we want to find the size of the other component C_{op}, we take S and multiply it by the ratio $R_{op}(\alpha)$. ($R_{op}= C_{op}/S$, so $S\times R_{op}= C_{op}$).

In the last diagram (α=53°) this ratio, i.e. $R_{op}(53)$, is approx. 0.8; S is 25, so C_{op} ought to be $25 \times 0.8 = 20$, which, again, it is.

There is a third ratio which could be (it was) calculated and tabulated for all α: the ratio between the two components (C_{op} /C)

(In fact, rather then measured, these ratios can be computed, to any degree of accuracy. How this is done is beyond the scope here, unfortunately so because it is a wonderful procedure).

So now we also have the answer for the second question raised at the beginning of this chapter: why it is important to have tables for *ratios* of the lengths of sides of triangles with a 90° angle.

9.3 Watch your tongue

Now that we have seen what these ratios are and what they do, it remains to give them names, but we *must* find alternatives for the taunting and arbitrary terms 'cosine', 'sine' and 'tangent'.

(The name of the subject itself, Trigonometry, from Greek for 'triangle measurement', can stand...)

(Reminder: an 'at90°' line is a line that is at 90° to some other line.)

We have a triangle with two lines at 90° to each other, and a third line - the longest one. For one of the two non-90° angles - which we call α – we want to define the ratios between the various pairs of lines in the triangle. Of the two at 90° lines, one is opposite to angle α and the other is next to α. A case could be made for using terms that divulge what we are talking about:

the longest line (which we know cannot be at 90°): 'LONG'

the at90° line opposite to angle α: 'OPP'

the at90° line next to angle α: 'NEXT'

Remember: if a line is not the LONG one, i.e. if it is OPP or NEXT, it must be an at 90° line.

(The (α) specifies the angle which the 'at90°' lines are said to be "opposite" or "next" to).

Lastly, "*n* divided by *m*" can be shortened to "*n* by *m*".

The three ratios then become:

Pronounced	Written (notation)	What the others, unhelpfully, call it
NEXT by LONG (α)	NX/L (α)	COSINE (α)
OPP by LONG (α)	OP/L (α)	SINE (α)
OPP by NEXT (α)	OP/NX (α)	TANGENT (α)

- one extra syllable, and so much misery spared...

(Again, this remedy is prescribed to be taken until one is fit in the way of comprehending the concepts. Thereafter one can abuse oneself with 'sines', 'cosines', whatever.)

Note: the ratio 'R' in the earlier example (refer to diagram) is now NX/L, and the 'R_{op}' is OP/L.

In the case of certain angles α we do not need tables to find ratios.

In the last chapter we found that if α = 45° the two at 90° lines are equal, so their ratio is 1,

- this means: OP/NX (45°) = 1 (ex. TANGENT (45°) = 1)

We also saw that if α = 30° the OPP is equal to half LONG,

- this means: OP/L (30°) = ½ (ex. SINE (30°) = ½)

9.4 From one relationship to another

There are simple connections between the ratios. What that means is that if only either of the first two ratios in the above table is known (for a given angle α) the other one can be worked out from it, and then the third can be calculated from the first two.

The way to get the third, OP/NX, from the first two, NX/L and OP/L, is fairly obvious: simply divide one by the other, i.e. $\frac{OP/L}{NX/L}$, the 'dividing by L' at the top and the bottom cancel each other out, leaving the desired $\frac{OP}{NX}$.

For example: in

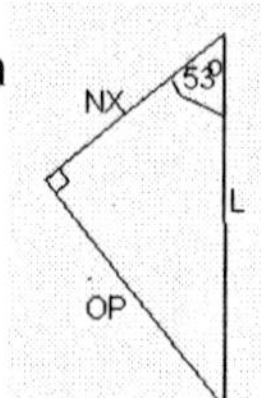

we found that OP/L(53) is (approx.) 0.8

and NX/L(53) is " 0.6,

so OP/NX(53) should be " $0.8/_{0.6}$ = 1.33,

which it is, as can (should) be checked in the drawing.

Note: not only is our OP/L, NX/L & OP/NX more revealing than "sine, cosine & tangent", but compare also the new and old forms of presenting the above connection *between* the three ratios:

Our notation *itself* suggests the reason for the above connection $OP/NX = \frac{OP/L}{NX/L}$

while not the slightest hint is gleamed from " $tangent = \frac{\text{sine}}{\cos ine}$ ".

9.5 What a wonderful idea this ancient came up with

To understand how the first and second of these ratios (OP/L & NX/L) are related we must begin by showing and proving one of the most striking relationships in maths: that very special relationship between the lengths of the three sides of any triangle in which one of the angles is 90°:

> If you measure the lengths of the sides and work out the squares of the lengths, you will find that the square of the long side equals the sum of the squares of the two sides which are at 90°.

For example, in a 90° triangle, if one of the 'at 90°' sides is 5cm and the other at 90° side is 12cm, the long side must be 13cm, because $5^2 + 12^2 = 13^2$.

This is easily verified by drawing two at 90° lines to these dimensions, 5 and 12 cm, and then measuring the distance between their free ends (the 'LONG'). Confirming this with every 90° triangle that one *tries* is reassuring, but not proof that it works with *any* 90° triangle. About one thousand years elapsed after the discovery of this remarkable relationship and the time when somebody found a way of proving it. Why?

As we have said earlier, in the realm of shapes it is often difficult to arrive at results by deduction alone. An *idea* is required, usually in the form of a diagram that is far from obvious. Mr. Pythagoras came up with one, 2500 years ago, and it is still one of the most elegant ideas ever conceived.

Naturally, the diagram could be expected to include a 90° triangle and some squares, but how should they be disposed?

This really belongs in the chapter on SHAPES. It is, however bad teaching practice to fuss over any topic before its significance has been made clear. In this case we needed to wait until we had found out why 90° triangles deserve so much attention in the first place).

This is the great idea for a diagram by which the relationship is proved:

a large square ABCD, with a smaller square of sides *l*, tilted so that its corners touch the sides of the large square and thus form four triangles (which are identical because the way the drawing is made up of squares you would not know if someone rotated it through 90° while you were not looking). They are 90° triangles, the long side having length *l* and the at 90° sides lengths k and n.

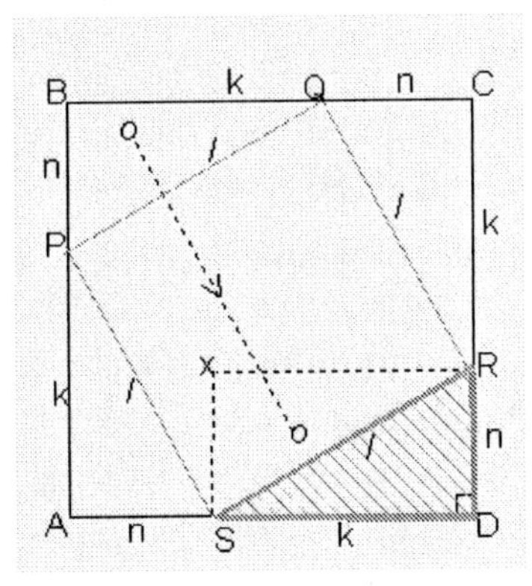

Looking at the larger square ABCD, we can see that the length of its sides is k + n. The area of the square is then $(k + n)^2$, which we know is $k^2 + n^2 + 2kn$.

This must be equal to the sum of the areas of all the shapes contained in the square: the smaller square PQRS, the area of which is l^2, and the four triangles, each the same area as triangle SRD.

If triangle PBQ is 'brought down' to join with SRD along their long sides they make a rectangle (XRDS) of area k × n, as do triangles SAP + QCR. The combined area of all four triangles thus amounts to 2kn.

Thus, equating the area of the big square with the total area of its contents,

$$\downarrow \qquad\qquad \downarrow$$
$$k^2 + n^2 + 2kn = l^2 + 2kn$$

The two sides remain equal if we remove the 2kn from both, leaving behind the

Pythagoras relationship $\mathbf{k^2 + n^2 = l^2}$.

Note: the shape of the triangle for which we proved this depends how long *l was* relative to the large square (the longer *l* is, the 'flatter' the triangle – check in the diagram). We did not specify what *l* was, so we proved it for 90° triangles of any shape.

We wanted to find a connection between OP/L and NX/L. Therefore we obviously need to know of some connections between OP, NX and L; that is what we have just found. Consider any of the (90°) triangles, for instance SRD. Depending on which of the two non-90° angles we choose to call α, one of the at 90° sides (k or n) would be the OP and the other the NX (*l* is of course the L(ONG), so the Pythagoras relationship can be written as

$$OP^2 + NX^2 = L^2$$

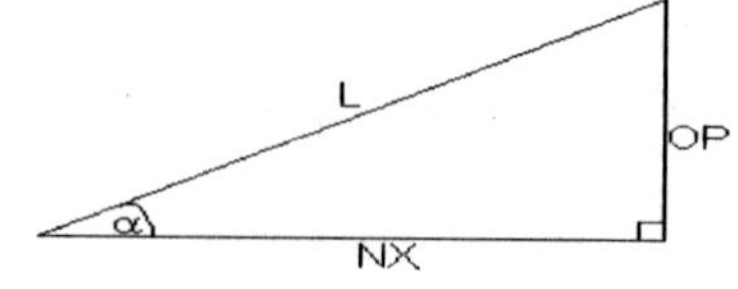

How can this lead to a connection between OP/L and NX/L? For one we want the OP and NX divided by L, but there is another issue: the Pythagoras relationship that we depend on deals with squares of the lengths. Getting rid of squares in equations is, as we already know, messy. Instead, we will satisfy ourselves with a relationship between the *squares* of OP/L and NX/L i.e. $(OP/L)^2$ and $(NX/L)^2$ which we will write (without the brackets) as OP^2/L^2 and NX^2/L^2 to bring it as close as we can to the constituents of $OP^2 + NX^2 = L^2$. To complete the match, namely to re-write this equation so that the required OP^2/L^2 and NX^2/L^2 appear, we obviously divide (both sides) of $OP^2 + NX^2 = L^2$ by L^2:

$$\frac{OP^2}{L^2} + \frac{NX^2}{L^2} = \frac{L^2}{L^2} \quad (=1)$$

or, bringing back the brackets:

$$(OP/L)^2 + (NX/L)^2 = 1 \qquad (\text{ex. sine}^2 + \text{cosine}^2 = 1)$$

We can try this out:

We had OP/L(53°) = 0.8 and NX/L(53°) = 0.6; indeed $0.8^2 + 0.6^2 = 1$.

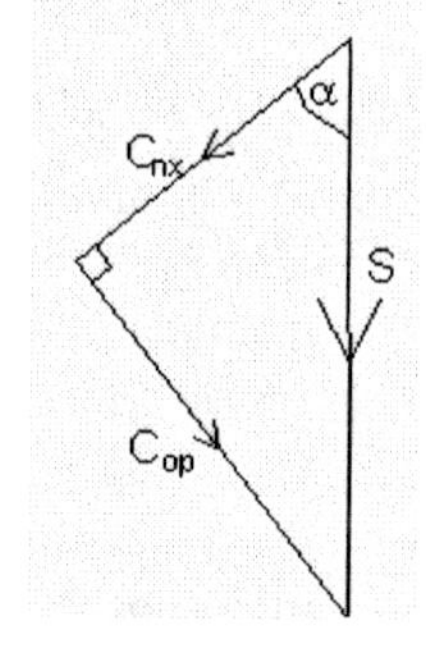

Let us return now to what all this was done for: working out the size of (mutually independent) components.

We had a 90° triangle, the long side representing a 'source activity' S, C_{nx} & C_{op} representing two mutually independent components, C_{nx} the one next to α and C_{op} opposite to it.

We had established that C_{nx} can be found from $S \times R$ where R is the ratio between the triangle's sides which represent C_{nx} and S.

We now call ratio R above (between component C and the source) "NX/L(α)", remembering that L and NX do not have to be equal to S and C_{nx} respectively, it is only the ***ratio*** between NX and L that matters and this depends only on angle α !

So, given a source activity of magnitude S, its component next to α:

$$C_{nx} = S \times NX/L(\alpha), \qquad \ldots S \times \cos(\alpha)$$

and the component opposite to α:

$$C_{op} = S \times OP/L(\alpha). \qquad \ldots S \times \sin(\alpha)$$

And wherefrom the values of NX/L(α) & OP/L(α) ? In the past from tables, now from calculators.

9.6 "Trig. ratios" everywhere

There is one trigonometric relationship that we all encounter *in the street*. For drivers who respect themselves in top gear only, or who drive around with no brakes, there are warning signs that look like:

This means that for every five units of length, for instance metres, that you cover along the road there is a vertical change in height of one metre:

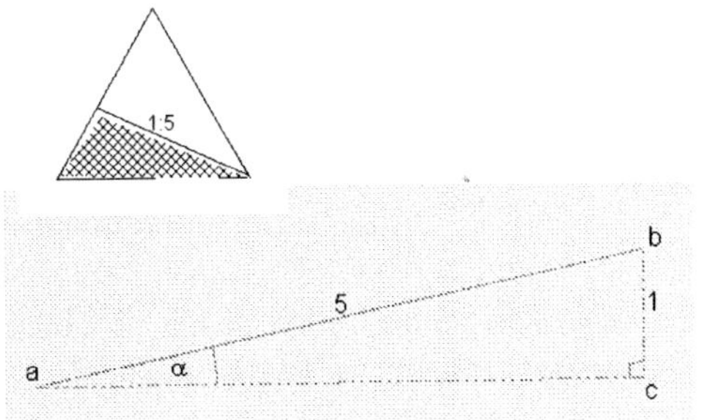

Where, might we ask, does the 90° come into the picture?

"To rise 1m for every 5m progress along the road" implies the following:

Imagine a *horizontal* (dotted) line leading from 'a' to the right. When it reaches the point that is situated *vertically* under b the distance between it (c) and b is 1m. Therefore the *horizontal* a-c and the *vertical* c-b lines form the at 90° lines of the 90° triangle. The LONG of the triangle is the road. The "angle of the slope" is of course the angle at 'a', which we call α. With respect to α the vertical line is the OPP and the horizontal line the NX. In this case the OPP is the 1m rise, and the LONG is the 5m progress along the road during which the elevation takes place.

Therefore what "1:5" slope means is that *the OP/L of the angle of the slope is 1/5.*

(Reference to the OP/L table will show that the angle is 11.5°.) OP/L –sin

Incidentally, estimating the angles of slopes is usually very misleading; the worst slope you might encounter on real roads, 1:4, makes you feel as though you are driving up a wall. In fact it is no more than

14.5°

9.7 Just one of a billion applications

To conclude this chapter let us look at a practical application of the relationships - all three of them - in a 90° triangle.

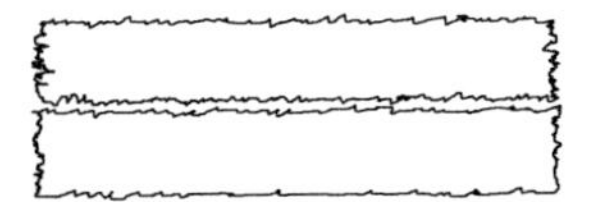

The microscopic texture of flat surfaces of solids, even of smooth ones, is rugged. When such a surface rests on another the peaks of one surface tend to interlock with the *valleys*, and or 'weld' with the *peaks* of the other surface (due to the high pressures concentrated on the thin peaks). Therefore when one attempts to shift one solid relative to the other along their interface the interacting peaks start pushing *sideways* on each other and so create a resistance to the shift. This resistance is called 'friction'; its direction is always such as to opposes the shifting force. The shifting force and the friction act, naturally, in the plane of the interface.

As long as no movement results from a gradually increasing shifting force it means that the opposing force – resulting from the interaction between the peaks (the friction) is always only strong as the shifting force, and grows equally with a growing shifting force. However, the friction resistance can only grow up to some maximum value, and if the shifting force increases further the shifted object "breaks free" and relative motion occurs. Measurement of friction usually refers to this maximal friction level.

As would be expected the strength of this maximal friction depends on how strongly the two surfaces are pressed together. In fact, with most materials the maximal friction is found to depend only on the total amount of force pressing the two surfaces together, irrespective of the size of the interfacing area: a cigar box would create the same friction irrespective of which face it rests on. The only other main factor is the nature of the materials of the interfacing surfaces.

Specifically, the maximum friction (*FR*) is equal to the 'press together' force (*PRS*) multiplied by a number which is always the same for a given pair of interfacing materials.

For instance, in highly frictional materials such as brake pads rubbing on steel, this number is approximately 0.5. This means that if these materials are pressed together with a force of 10 Kg, a 'shifting force' (*SHF*) of 5 Kg is needed to cause one surface to start sliding on the other (or, in other words, a maximal force of 5kg could be braked). With very friction*less* materials, such as Teflon, this number might be as low as 0.05. This means that, given the same 10 Kg 'press together', a shifting force of as little as ½ Kg would cause a slide.

This 'material constant' number ought logically to be called '*multiplier for friction*'; the traditional term is '*coefficient of friction*' and it is (therefore?!) written 'μ'.

Let us look at an illustration of how these forces interact.

A block 'B' rests on a slope at α° to the horizon. Let us examine why the block begins to slide as the 'steepness'(α°) is increased and at what α° the slide commences, given the 'μ' for the materials of the block and the slope.

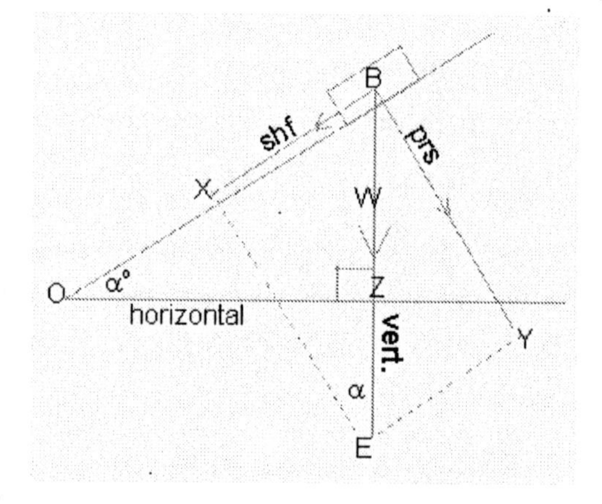

The source of all the forces involved here is the weight of the block W, that is, the force of gravity exerted vertically downwards on the block. This force 'W' is the source for the following:

- The 'shifting force' (*SHF*) along O-B.
- Pressing the block onto the surface of the slope (*PRS*). The direction of this ('pressing against') is of course at 90° to the interface on slope O-B. This also means that *PRS* and *SHF* are at 90° to each other, and are therefore *mutually independent components* of W.

SHF attempts to shift the block down the slope while the frictional force tries to prevent this. The latter depends of course on the *PRS*. The steeper the slope angle α, the greater the tendency to shift (*SHF*) and the smaller the friction-inducing 'press-together' force (*PRS*). The question is:

- at what angle will *SHF* exceed the maximal value of the friction, namely, $PRS \times \mu$?

Naturally this depends on how large *SHF* and *PRS* are in relation to their 'source' W, given α. Such a question leads our attention to a triangle whose sides represent these forces: BXE, bearing in mind that *PRS* can be represented by line XE because it is equal (in direction and length) to BY.

First though we must also establish that angle XEB is equal to α; that is to say, that ∡XEB =∡ZOB. (Diagram!) These are matching angles of triangles XEB and ZOB. These triangles have ∡XBE in common and both have a 90° angle, ∡BXE and ∡BZO respectively. So, their third angles, ∡XEB and ∡ZOB, must also be equal.

So, we focus our attention on triangle BXE, shown here enlarged:

As angle BXE is 90°, w is represented by the 'LONG' (L),

shf " " " " 'OPP' (to α) and

prs " " " " 'NEXT' (NX).

As we already know, we can get *SHF* by multiplying *W* with the ratio between the two, namely

$$SHF = W \times \frac{SHF}{W}$$

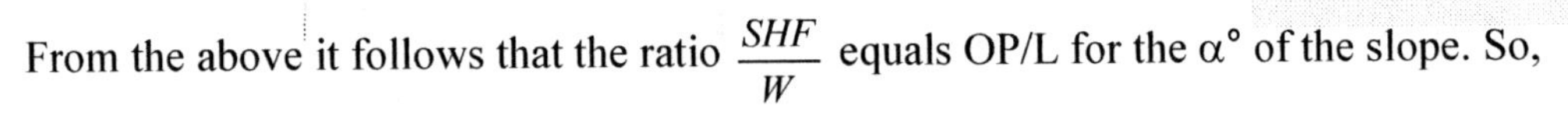

From the above it follows that the ratio $\frac{SHF}{W}$ equals OP/L for the α° of the slope. So,

$SHF = W \times$ OP/L (α) ... $W \times \sin(\alpha)$

Similarly, $PRS = W \times$ NX/L (α) ... $W \times \cos(\alpha)$

For example, if W is 5 Kg, drawn above at the scale of 1 cm to 1 Kg, and α is 24°, looking up the *OPP*by*LONG* tables shows that (approx.):

OP/L(24°) = 0.4, so $SHF = 5 \times 0.4 = 2$.

and the *NEXT*by*LONG* table shows that

NX/L(24°) = 0.91, so $PRS = 5 \times 0.91 = 4.55$

(which we can/want/must check against the drawing.)

For small slope angles we expect the shifting force to be small. Indeed, for small angles their opposite at 90° line is small relative to the LONG i.e. small angles have a small OP/L and so the *SHF* component of *W* (equal to $W \times$ OP/L (α)) is small too.

We remember that at every stage, starting from a shallow slope, the friction is only big enough to resist the shift, so, as the angle grows and the shifting force increases, so does the friction resistance. The latter however reaches a maximum level which equals $PRS \times \mu$, and also the size of *PRS* itself depends on the slope angle α ($PRS = W \times$ NX/L (α) - this one *decreases* with increasing α).

Therefore the block breaks free and starts sliding when the slope angle α reaches a value at which

SHF equals *PRS* × μ

which means: $W \times$ OP/L (α) = $W \times$ NX/L (α) × μ

Both sides can be divided by *W*, leaving OP/L (α) = NX/L (α) × μ

This, incidentally, means that the 'slip' angle does not depend on the weight of the block!

Dividing both sides by NX/L (α) we get $\frac{OP/L(\alpha)}{NX/L(\alpha)} = \mu.$

We already know that $\frac{OP/L}{NX/L}$ is *OP/NX*, so the above means *OP/NX*(α) = μ ...tan(α) = μ

If the "coefficient of friction" is, for example, 0.3, the block will begin to slide when the slope is raised to the angle of which the *OPP*by*NEXT* is 0.3. We find this angle, by searching through the OP/NX table, ('tan' table) to be 16.7°.

This method can be used to determine the value of μ for any pair of materials. All that need be done is to place a block of one material on a plane made of the other material; tilt the plane progressively until the block begins to slide; measure the angle at which this happens, and look up the *OPP*by*NEXT* for this angle.

We can now summarise the main purpose of the "trigonometric ratios":

In a 90° triangle with one of the (non 90) angles known, given the length of any one side, these ratios enable us to work out the length of the other sides. Let us run through all the combinations: (it is now clear from the diagram which side is *OP, NX & L*)

If **c** is given and we want **a**, we get it, as we know, from **c** × OP/L(α) (because OP/L here is a/c), but what do we do if **a** is given and we want **c?**

The safest general way is to start by selecting the ('trig') ratio that deals with the given and the wanted sides, in this case it is OP/L(α), then write down what it means using the symbols chosen to represent the sides (here **a** & **c**):

$$\text{OP/L}(\alpha) = \frac{\mathbf{a}}{\mathbf{c}}$$

and then proceed with the necessary steps that end up with the required item standing alone in front of the = sign. In this case, where we want "**c** = ", we multiply (both sides, of course) by **c** – this gets a '**c**' up in front of the = on the left, and then divide (both sides) by OP/L(α) to leave the **c** alone, ending up with

$$\mathbf{c} = \frac{\mathbf{a}}{OP/L(\alpha)}$$

An irresistible exercise should be to verify also the connection between the other two pairs, **b** & **c** and **a** & **b**:

$$\mathbf{b} = \mathbf{c} \times \text{NX/L}(\alpha) \quad \text{or} \quad \mathbf{c} = \frac{\mathbf{b}}{NX/L(\alpha)} \qquad \text{and} \qquad \mathbf{a} = \mathbf{b} \times \text{OP/NX}(\alpha) \quad \text{or} \quad \mathbf{b} = \frac{\mathbf{a}}{OP/NX(\alpha)}$$

Aliens-communication code
OP/L –sin NX/L –cos OP/NX -tan

9.8 DIY tables

This was to have been the end, but as it is so much fun let us find the values of some more 'trigonometric relationships' for special angles where we can work it out instead of looking up in tables. We have already seen two such cases: OP/NX(45°) and OP/L(30°).

Before looking at these special angles there is a bonus worth noting. Having worked out a relationship for an angle, we can use the result to determine another relationship of another angle:

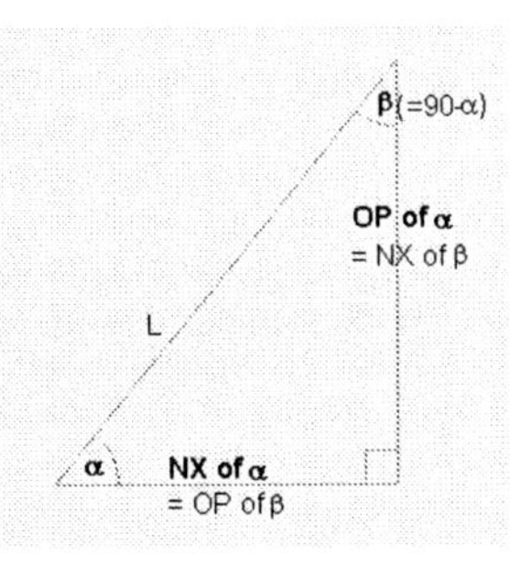

When a side is OPP to one of the two (non 90°) angles (eg. angle β), that side is NEXT to the other (non 90°) angle (e.g. angle α).

herefore NX/L(α) is the same as OP/L(β), and vice versa.

Also, by carefully following who is who in the diagram and remembering ('crushing ladders') that

$$a/b = \frac{1}{b/a}$$

we find that $\quad OP/NX(\alpha) = NX/OP(\beta) = \dfrac{1}{OP/NX(\beta)}$, and vice versa.

Now, the sum of the three angles α, β & 90 is 180, so α+β must be 90, which means **β = 90-α** (and equally α = 90-β)

If we replace β with this 90-α in the above, we get

$$OP/NX(\alpha) = \frac{1}{OP/NX(90-\alpha)} \qquad \ldots \tan(\alpha) = \frac{1}{\tan(90-\alpha)}$$

and

$$OP/L(\alpha) = NX/L(90-\alpha) \qquad \ldots \sin(\alpha) = \cos(90-\alpha)$$

For instance, we already know that OP/L(30°) = 1/2 ...sin(30) = 1/2

therefore we also know now that NX/L(60°) = 1/2 ...cos(60) = 1/2

Now let us turn to the special angles mentioned above, for which relationships can be deduced from simple, known properties of the shapes, that is, with angles of 0°, 30°, 45°, 60° and 90°.

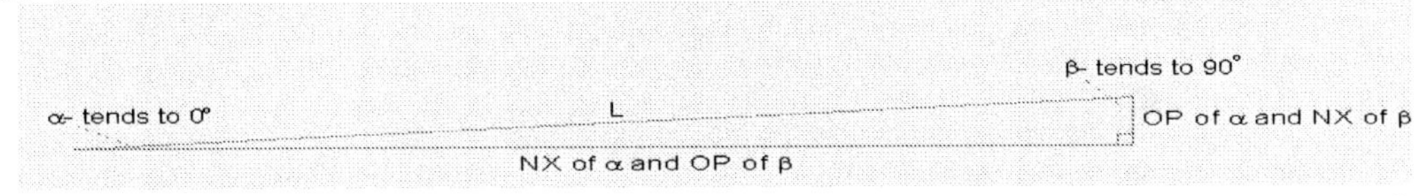

First let us look at a (90°) triangle in which α = 0°. This cannot be drawn, of course, so the triangle above is drawn in such a way that α is *nearly* 0; its OPP is very small, and when angle α closes up completely (α= 0), its OPP also shrinks to 0.

0 divided by anything, such as 0/L or 0/NX results in just as round a 0. Therefore

OP/L(0°) = 0, and OP/NX(0°) = 0. ...sin(0) = 0 = tan(0)

It is also clearly seen that NX of α become equal to L as α tends to 0, so we also know that

NX/L(0°) = 1 ...cos(0) = 1)

To get the ratios for 90° we can use the same diagram but look at it 'from *β*'s point of view', (*β* = 90-α, so it tends to 90 as α tends to 0): now the *OP* (of *β*) becomes equal to L, so

OP/L(90°) = 1 ... sin(90) = 1

and the *NX* (of *β*) tends to 0, so

NX/L(90°) = 0 ...cos(90) = 0

But what happens to OP/NX when the angle (β in the diagram) becomes 90°?

NX (of β) becomes 0, and so a healthy OP gets divided by 0 (wish this happening to your capital): we know that anything divided by 0 yields infinity. So

$$\text{OPP/NX}(90°) = \infty \qquad (\tan(90) = \infty)$$

Next let us look again at the 30°/60°/ 90° triangle.

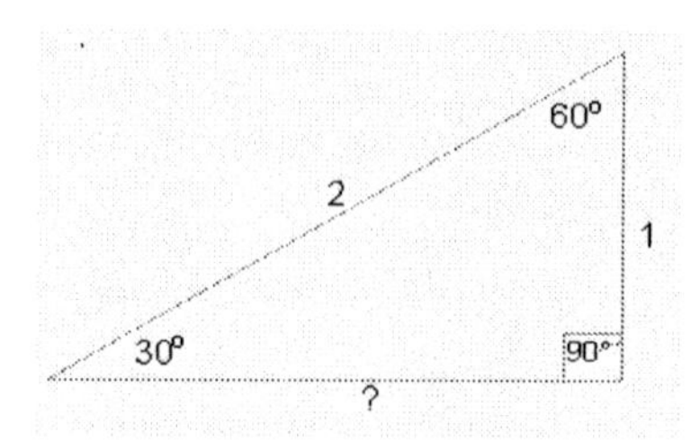

We need the relative lengths of the lines of this triangle. We know that the length of the at90° line opposite the 30° angle is half the length of the LONG. In the diagram we can mark these two lines with lengths 1 and 2 respectively (any numbers with this ratio will do, so we choose the simplest). How long is the other at90° line?

As we want to find the length of a side in a 90° triangle, given the lengths of the other two, we use a relationship that is well known to us and to Pythagoras:

applied to the sides in the diagram, it is $1^2 + ?^2 = 2^2$

i.e. $1 + ?^2 = 4.$

Fortunately we love solving equations. We need ? = (numbers only)

We begin by getting rid of the 1 by subtracting it from there (and also from the other side), leaving

$$?^2 = 3$$

The ^{2} goes by taking the 2-way fragment ("root") on both sides, leaving

$$? = 3^{1/2} \quad (\text{ex.} \sqrt{3}\ )$$
$$= 1.732...$$

Thus the relative lengths of the 30°/60°/90° triangle are:

Paying ***careful attention*** to which side is OPP & NX to which angle we find that

$$\text{NX/L}(30°) = 3^{1/2} / 2$$
$$= 0.866... \text{ which is the same for OP/L}(60°),$$

OP/NX(30°) = $1/3^{1/2}$ (A clever 'trick' for simplifying this will be shown below.)

and OP/NX(60°) = $3^{1/2} / 1$ (i.e. 1.732...)

Aliens-communication code
OP/L –sin NX/L –cos OP/NX -tan

Finally we work out ratios for 45°. If one of the non-90 angles is 45 so is the other (because together with the 90 they all add up to 180). So, we look at the 45°/45°/90° triangle:

We know that the two 45 angles make it a 2-equisided triangle: opposite the equal angles there are equal sides, let us say of length 1. (That was the reason for OP/NX(45°) being 1/1, or 1).

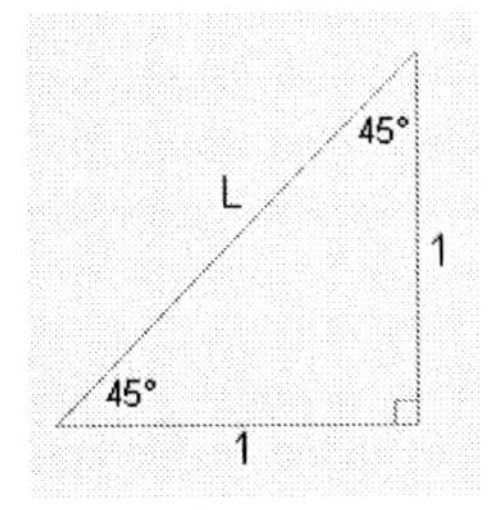

But what is L, which we need to know in order to work out the OP/L and NX/L for 45°?

We Pythagoras addicts know that in this case $L^2 = 1^2 + 1^2$, which means $L^2 = 2$.

Again we rid L^2 of the ^{2} by taking the 2-way fragment on both sides: $L = 2^{1/2}$ (= 1.414...)

So the relative lengths of this 45°/45°/90° triangle are:

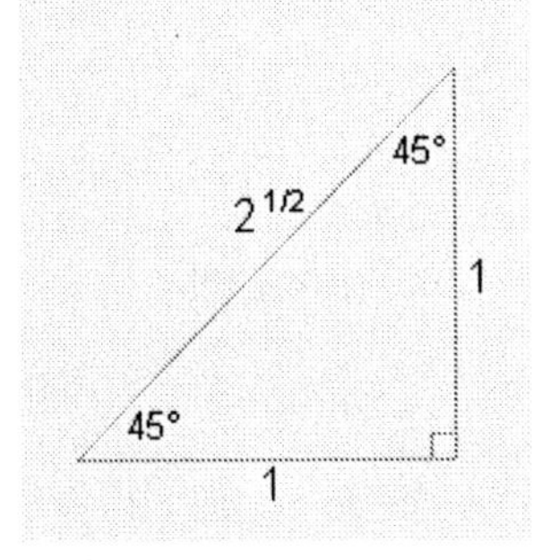

Which means that $OP/L(45) = 1 / 2^{1/2}$, and NX/L(45) is of course the same thing.

Now we come to a neat trick, a very clever idea of whoever first had it.

It has to do with the fact that $3^{1/2}$ and $2^{1/2}$, which were involved in the triangles above, have an endless string of digits behind the point: 1.73205 etc. etc. and 1.41421.... When we only want to write them for their own sake we can just go as far in the series of digits as the accuracy of the answer demands. The same applies when we need to divide them. But when we want to divide *by* such a number it is not so clear how many digits must be included in the divider in order to achieve the desired accuracy of the result. And if, heaven forbid, we have to do it manually by long division, it is very inconvenient to have such a long divider. How then do we manage, for instance in the case of :

OP/NX(30°), which is $1/3^{1/2}$ (= 1/1.73205...), and of

OP/L(45°) (and NX/L), which is $1/2^{1/2}$ (= 1/1.41421...)?

For the $\frac{1}{3^{1/2}}$ we can simply multiply top and bottom by $3^{1/2}$, i.e. : $\frac{1 \times 3^{1/2}}{3^{1/2} \times 3^{1/2}}$. The value, of course, is unchanged, but the nuisance divider $3^{1/2}$ now becomes $3^{1/2} \times 3^{1/2}$, a neat 3 (!)

Thus in OP/NX(30°), instead of $\frac{1}{3^{1/2}}$ we now have $(\frac{1 \times 3^{1/2}}{3^{1/2} \times 3^{1/2}} =) \frac{3^{1/2}}{3}$, i.e. $\frac{1.732...}{3} = 0.577...$

Much easier to do, and we know how many digits to divide: as many as needed in the answer.

Similarly:

- for OP/L(45°), instead of $\frac{1}{2^{1/2}}$ we now have $(\frac{1 \times 2^{1/2}}{2^{1/2} \times 2^{1/2}} =) \frac{2^{1/2}}{2}$, i.e. $\frac{1.414...}{2} = 0.707...$

Now we have enough to make up a table, all of our own doing:

Angle α	OPP by LONG OP/L (= "sine")	NEXT by LONG NX/L (= "cosine")	OPP by NEXT OP/NX (= "tangent")
0	0	1	0
30	$1/2 = 0.5$	$3^{1/2}/2 = 0.866...$	$1/3^{1/2}$ $(=3^{1/2}/3)$ $= 0.577...$
45	$1/2^{1/2}$ $(=2^{1/2}/2)$ $= 0.707...$	$1/2^{1/2}$ $(=2^{1/2}/2)$ $= 0.707...$	1
60	$3^{1/2}/2 = 0.866...$	$1/2 = 0.5$	$3^{1/2}/1 = 1.732...$
90	1	0	∞

This drawing shows 90° triangles with various angles α, all located at o; in all of them the LONG is of the same length (this is why their tips trace a circle), OP are the vertical lines and NX are horizontal.

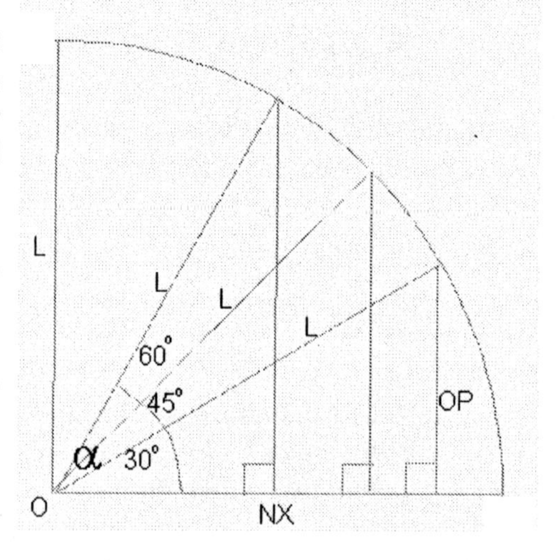

If we examine the table and the drawing we notice that:

OP/L and NX/L (sin & cos) are never more than 1, because the divider is the LONGest.

As α grows from 0° to 90° OP/L grows from 0 to 1 (because OP grows until it reaches L),

however NX/L goes *down* from 1 to 0, because NX gets smaller.

As for OP/NX (tan), it starts at 0, reaches 1 at α=45° and then just goes berserk (because OP grows *and* NX gets smaller, ultimately becoming the explosive divider 0).

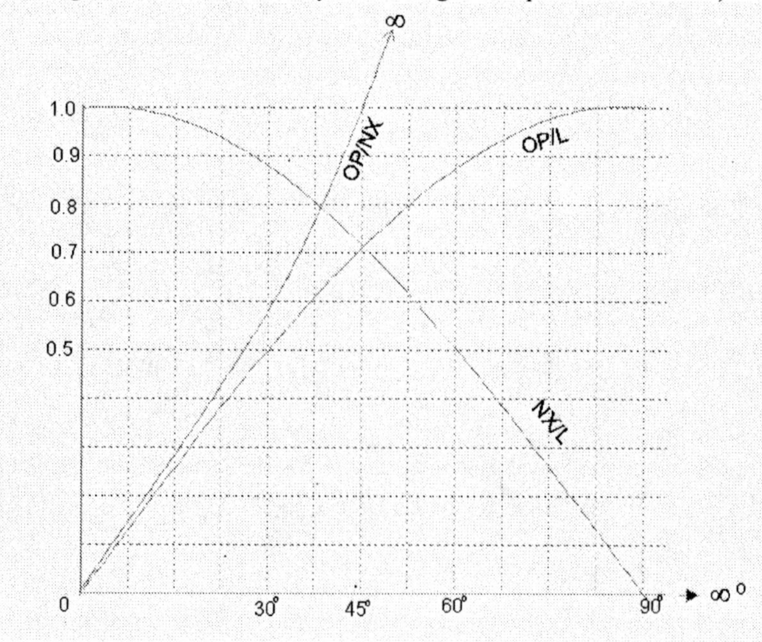

Chapter 10

Steepness

When we say that factors such as 'time' and 'height' are *independent* (of each other) we need to clarify what we mean by this, because there *could* be situations in which they are *not* independent: think, for instance, of water level in a leaking tank.

However, when there *are* connections between *independent factors*, this occurs only in specially defined situations such as the above example. In general though the passage of time does not *necessarily* entail change in height, and *vice versa*. This contrasts with, for instance, 'time' and 'age' which *are* dependent: you cannot retain the latter while the former runs on, no matter what the advertisements say. The same might even be said of 'elegance' and 'simplicity': they are not the same thing, but they are always bound to be connected to a certain extent, because nothing is elegant without a measure of simplicity.

Thus 'independent factors' are ones which *can* be unconnected; ones that are not *bound* to be connected by virtue of a similar inherent meaning. When they *are* connected it is only in special cases in which there is separate information about a connection, information that cannot be deduced from the definitions of the factors alone.

10.1 Describing connections

There are various ways in which such connections can be stated.

Let us take distance and time as an example. These are independent concepts: the fact that the disco is nearer than the real zoo does not mean that it is later in one than in the other. However, in the *special case* of driving along a straight road, distance from the starting point does depend very much on what time it is in relation to the time the driver set off. To establish a *quantitative* connection further information is required, for example "constant speed of 60 miles per hour". The connection could be presented in four different ways.

I. In a TABLE, that is, a list of different times, and next to each, the distance from the starting point at which the driver was at that time, for instance

TIME (hours from starting time)	DISTANCE (miles from starting point)
0	0
1	60
2	120
2.5	150
4	240

II. In a GRAPH, in the form of a line (or curve) drawn relative to a point O called 'origin', where each point on the line represents *both* a time and (the corresponding) distance. How?

The horizontal (sideways) displacement of such a point relative to the 'origin' represents time elapsed since the start, at a scale of, for instance, 1 cm on the paper per ½ hour. The vertical (upward) displacement of the same point relative to the origin represents the distance covered in that time, again on a scale, for instance of 1 cm on the paper per 60 miles. In order to establish the 'sideways' and 'upward' displacements relative to the origin two lines are drawn first:

a horizontal line going through the 'origin', called the 'time axis' because 'moving' along it - *or parallel to it* - represents changing the time, but without changing the location on the road - all the points on this time axis represent location 0.

a vertical line, going through the 'origin', called the 'distance axis' because 'moving' along it - *or parallel to it* - represents changing the location on the road, but without changing the time - all the points on this distance axis represent time 0.

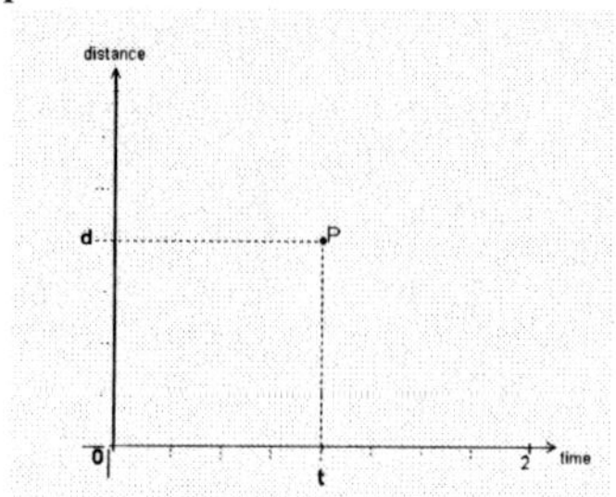

The time axis is marked with graduations showing

- hours, say, starting with 0 at the origin;

- the distance axis is marked with graduations showing miles, say, starting with 0 at the origin.

We said that 'P' represents both a time and a distance: the time is represented by P's sideways displacement from O. As the two axes are at 90° to each other we can determine this sideways displacement by dropping a vertical line from 'P' to meet the time axis at a point which we call 't'. The sideways displacement of point P from the origin is then equal to length O-t along the (horizontal) time axis. So the time represented by P can then be read off the graduations on the time axis, at t.

Similarly we can establish the road distance represented by P by drawing a horizontal line left from P to meet the distance axis at a point which we call 'd'. The position of d up the graduations on the (vertical) distance axis provides the road position of P.

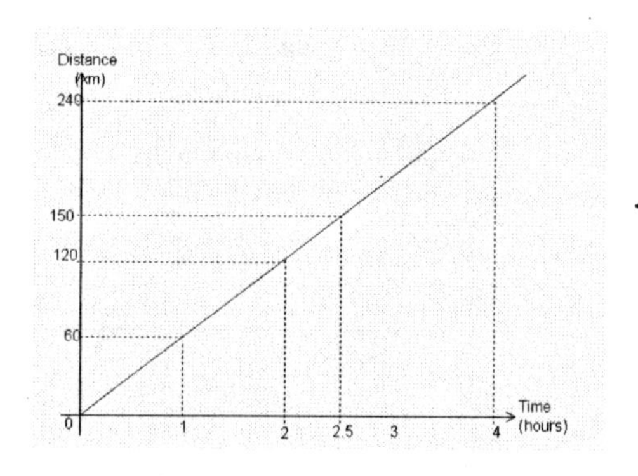

Each point on the following graph presents a certain time and the car's position along the road at that time. Specifically, this graph states the same connections between time and distance as are given in the table above, each marked dot on the graph corresponding to one of the items in the table.

(The points that make up the graph between the marked dots represent the car's interim position/time pairs.)

III. A verbal recipe for finding the distance (in miles, say) when the time (measured in hours) is given, i.e. a 'rule'. In this case the rule is simply "multiply the time (in hours) by 60".

IV. Applying our familiar shorthand to the above rule. This is the neatest way of presenting the connections: use d for "distance" (in miles), t for "time" (in hours), and shorthand the above 'recipe' to $d = 60 \times t$.

What has not been shortened is the long list of names that have been coined for this type of shorthand. These include '*expression*', '*formula*' and '*function*'. The present example would normally be referred to as "d, as a function of t", shorthanded: d = f (t). ('Function of' means here "dependent on...". Remember: this 'dependence' between these otherwise independent factors exists only for such special cases as here where one *chooses* to connect between them by way of some special 'story').

d and t here are called 'variables', because they may take on various numerical values.

The *function* $A = l^2$ tells us that something called 'A' for short depends on something else, called 'l' for short, and states that if, for instance, l grows five-fold 'A' will grow twenty-five-fold. This might, for example, represent an area A of a square which depends on its length (and width) l.

The function $t = \left(\frac{443-h}{4.91}\right)^{1/2}$ (ex. $\sqrt{\frac{443-h}{4.91}}$) tells you how much time (in seconds) after jumping off the 443m high Sears Tower (don't try this at home – not tall enough), you reach a height of h(metres).

For example:

you will be at height (h) of 320m at time $t = \left(\frac{443-320}{4.91}\right)^{1/2}$

$$= \left(\frac{123}{4.91}\right)^{1/2} = 25^{1/2} = 5\text{sec (after take-offf).}$$

With h = 0, the above function tells you when you will be no more.

The same connection between h and t can be provided by the function $\mathbf{h = 443 - 4.91\ t^2}$ which tells the 'reverse' story, namely, what height you reach when you are t sec into your downward journey up to Heaven.

Again this connection between height and time can be presented in the form of a graph:

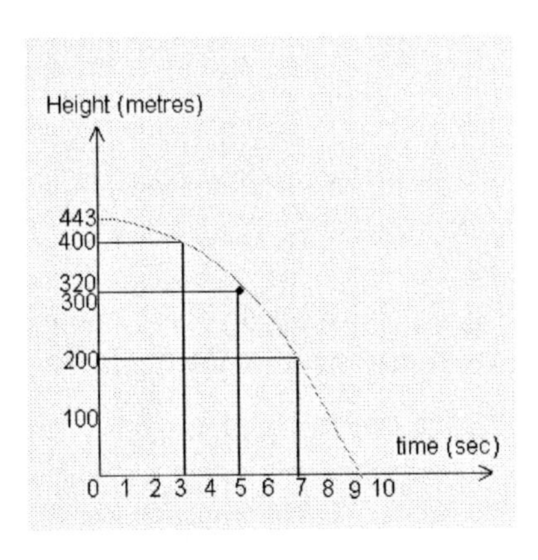

The vertical position of each point relative to the origin represents your distance from the ground at a time that is represented by the *horizontal* shift of the point (relative to the origin). For instance, point p states that 5 sec into the flight you are 320m above ground. (From there, your life expectancy is 4.5sec.)

Since a 'function' states the manner in which one variable depends on another variable, we call the first variable a 'dependent' variable and the second 'independent' or 'free' variable. The reason for these terms is that in, for instance, $d = 4u^2 - 5/u$, we can choose any value we like for u, whereas the value of d will be forced (to be the result of $4u^2 - 5/u$).

The dependent variable can depend on *several* 'free' variables. For example, the volume (V) of a rectangular container is a function of its depth (d), width (w) and height (h). In shorthand this is written $V = f(d, w, h)$. Specifically the dependence is $V = d \times w \times h$. In other words

$V = f(d, w, h)$ states *what* V depends on;

$V = d \times w \times h$ states *how* V depends on them.

Summarising: you *choose* a value for the free variable and perform on it the calculations which are prescribed by the "function" (the "arithmetic recipe"). The result is the corresponding value of the dependent variable. In short: a *Function* provides the value of the dependent variable, given the value of the free variable(s).

When we talk about "the *value* of a function" we mean the value of the dependent variable.

10.2 Things are improving rapidly. Don't care how bad they are

There are times when what we are interested in about the (dependent) variable is not its value, but something else connected with it: the rate at which it changes.

For example, we can read comfortably under a very wide range of illumination; bright sunlight produces nearly a thousand times more light than the illumination specified for libraries. Therefore the level of illumination does not matter much to the eye; however relatively small changes are easily detected if they occur fast enough: a cloud passing over the sun is noticed at once; someone playing with a dimmer switch is particularly irritating, and a flash can shock.

That is to say that when the dependent variable is strength of illumination, this strength varying with time (the 'free' variable), it is not the *size* of this dependent variable that matters to the eye, but how *fast it changes*, in other words the '*rate of change*'.

Many other examples could be given.

An employee who is in a junior position but is promoted every few months probably derives more satisfaction than a senior director who only accedes to a higher rank once in ten years.

Mains electricity does not come from batteries. It is generated by *changing* the strength or quantity of a magnetic field in the vicinity of a coil of conducting wire. It does not matter to the wire whether the magnetic field is tiny or enormously powerful; neither will cause electricity to flow. Only if and while the magnetic field is *changing* does electric current flow. If the magnetic field varies slowly a little current will be produced; if varied quickly a lot of current will flow. (If you know *why* this happens you will become very famous because no one else knows; this just happens to be a property of electricity and magnetism.) Thus, when it comes to lifting scrap metal with a magnet on a crane it is the *strength* of a magnetic field that counts, but only its *rate of change* matters in generating electricity.

What exactly is the difference between *change* and *rate of change*?

Let us take two cases of rise in temperature. In the first case temperature rose 9 degrees in six seconds. In the second the temperature only rose 6 degrees, but it did so in 2 seconds. The *change* in the first case was greater, but what of the *rate of change*?

What this question poses is "by how many degrees did the temperature rise *in one unit of time*?"

The rise of 9 degrees occurred within the space of 6 seconds, so in 1 second the rise was 1.5 degrees. In other words: the *rate* of change is 1.5 degrees per second. In the second case, although the total change was less, the rate at which it occurred was twice as great: 3 degrees per second. (We assume here that the heating occurs at a constant rate.)

If we represent these connections between temperature and time in the form of a graph we get:

(Since we are concerned only with *changes* in temperature it does not matter whether the 9 degree rise is drawn on the scale between 0 and 9, 1 and 10 etc.. Similarly, since it is the *duration* that matters, the 6 second rise can likewise span 0 to 6, 2 to 8 or any other six second period on the scale.)

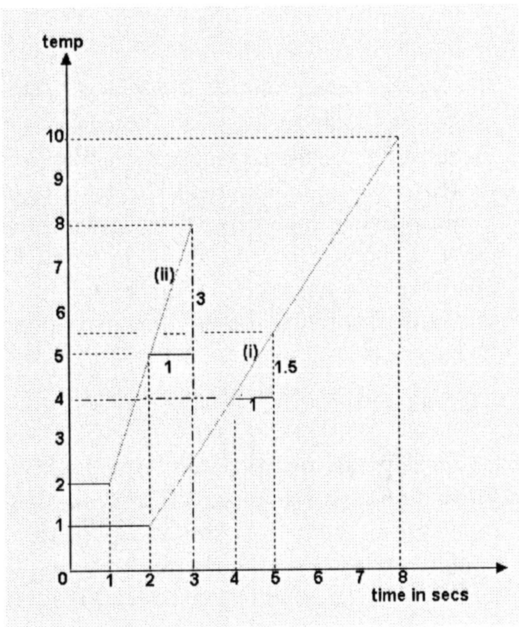

On each of the two graphs we can mark what happened during any *one second* (which is represented by one horizontal unit):

in **(i)** the corresponding temperature change is 1.5 degrees, (1.5 units on the vertical scale),

in **(ii)** the corresponding temperature change is 3 degrees, (3 units on the vertical scale)

Temperature here is the 'dependent' variable, because its level depends on what time point one chooses to check the temperature. 'Time" thus is the 'free' variable.

In this graph, which shows the dependent variable vertically and the free variable horizontally, the rates of change appear as slopes; the greater the rate of change the steeper the slope!

In the above example, of functions whose graphs are straight lines (and hence called 'linear functions'), the slope is of course the same everywhere. Functions however need not always be such that their graphs are straight lines.

An example of this is the graph of a function describing the height, at different times, of a stone thrown up in the air. We now know that the *slope* of the graph represents the *rate of change in height*, or how much height changes in each unit of time. Incidentally, this phrase, how much a position changes during one unit of time, has a name: *speed.* And *rate of change in height* simply means *vertical speed.*

The stone we throw up in the air initially (a) gains height rapidly, so the slope of the line describing its height as a function of time is steep. As the stone slows (b) due to gravity it gains less and less height per unit of time, so the slope becomes less steep. At some point (c) the stone comes momentarily to a halt before it begins to fall. In that instant the stone does not change height at all, so the line on the graph becomes horizontal, because time, however short, passes when the stone is stationary; the slope here is zero. From then on (d) height *decreases* in every unit of time, in other words the change is negative. What does this mean?

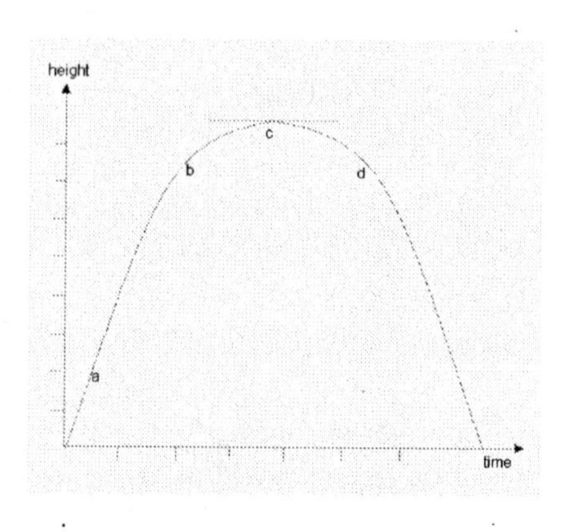

The exact meaning of '*change*' (of height, say) is: "later height minus earlier height". If the new height is *less* than the earlier one the result of the subtraction is negative. This negativity is also reflected in the graph by a downward direction of the slope. ("Upward" and "downward" relates to tracing the curve from left to right, that is, with time moving forward). "Downward" is the opposite of "upward", and we now know that "opposite" is represented in maths by a change of sign from + to -. Therefore on the way up we measure 'positive slopes', and on the way down, (the time continues to increase), we measure the 'negative slopes'.

What is the temperature outside?

That depends on what time you look at the thermometer, so this too is a 'function of time', and the rate at which it changes is not fixed either. Temperatures rise and fall, so the rate of change is sometimes positive and sometimes negative, and the rate is not constant: when the sun rises temperature changes rapidly, in other words the rate of change has a large value; later in the day temperature changes more slowly, so the rate of change has a lower value. The rate may even be zero, at periods when there is no change in temperature.

So, not only is there a different temperature at various times of the day, but the *rate* at which the temperature changes *also* varies with time. This means that also *the rate of change* of temperature is a function of time, and so it too can be presented in the form of a graph: the height of each point represents how fast the temperature changes, and the horizontal position of the point shows at what time it occurs. Here is how such a graph can be generated.

First we draw a graph of the temperatures at different hours. We have seen that when representing a function as a graph, the rate of change at different times shows as the slope of the graph at these respective times.

We then draw another graph underneath, using the same time scale, that is, the horizontal position of points represents the same hour in both graphs. The height of each point in the lower graph represents the magnitude of the *slope* of the temperature graph at the same hour, that is, immediately above. The higher the point in the new 'slope-size' graph the greater the slope on the temperature graph above it.

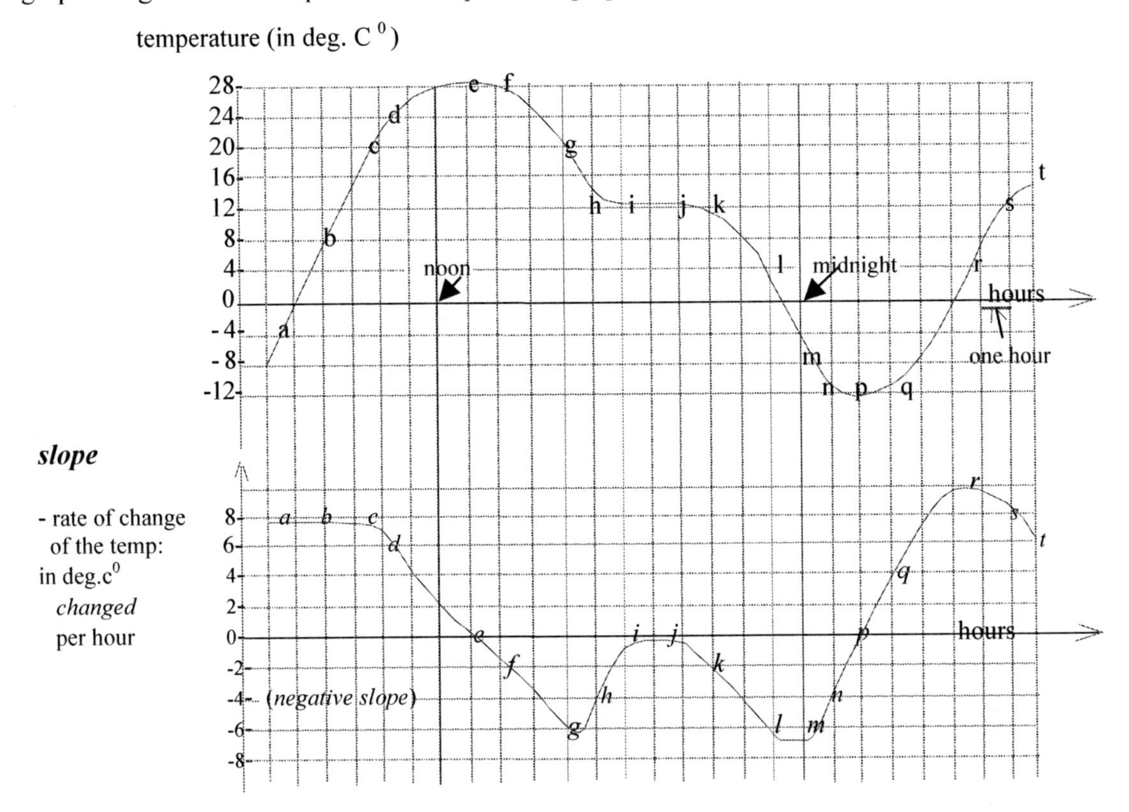

Perhaps this is not the prettiest of pictures, but the following notes will show how easy it is to understand what goes on up there. It is *vital* to go through this patiently. Having done so, you will begin to really understand what so-called "differential calculus" is about!

The temperature graph at 'a' is quite steep, that is, the slope at this point is relatively large, so its corresponding point '*a*' on the lower graph is high. (*Note*: we use a,b,c for the temperature graph and *a,b,c* for the ***slope*** graph). The slope at points b and c on the temperature graph does not change much, so *b* and *c* on the slope graph remain at the same level as *a*. At d however the slope becomes less steep, so the point *d* on the slope graph is lower than *a, b, c*. At e the temperature curve goes neither up nor down: the slope is zero, so the corresponding point *e* below is at (height) 0. From here on it is all downhill: this is the 'opposite', or 'negative' slope. At f the negative slope is moderate, so *f* is a little below the 0. The 'opposite' slope becomes steeper, in fact steepest in this section, at g, so *g* is, locally, most negative, that is, furthest below the 0 level. Near h this ***downward*** slope again gets less steep, and thus *h* ***rises*** again towards 0. From i to j things are level again, which means zero slope, so this is reflected on the slope graph by points *i,j* at 0. Consider sections e-f-g and j-l-k: although j,k,l is lower than e,f,g, their 'slope-story' is the same. This is why *e,f,g* and *j,k,l* in the slope graph are the same, at the *same level*. Between l and m the temperature graph has a large (negative) *constant* slope, so all the points between *l* and *m* on the slope graph have the same (large, negative) value. As the negative temperature slope becomes less steep at n, *n* rises towards 0, and reaches it at *p* because the temperature graph at p is level again. But it is only level at that precise point p, as the slope changes from a negative to a positive one, so the slope graph it only *passes through* the zero line at one *p*. Although q itself is still in the negative temperature range its slope begins to be positive again, so *q* is now above board, and is even more (most) so at *r* because the temp graph becomes steepest at r. At s and t it loses some of its steepness, and so at *s* and *t* the slope graph exits with a small bow. Bravo.

If, instead of temperature, the upper graph portrays the time-varying strength of a magnetic field surrounded by a coil of wire (with units of time in milliseconds rather than hours), the lower graph will show the amount of current flowing through the wire at the same times. This is because the current does not result from the *level of strength* of the magnetic field, but from the *rate* at which that level changes, in other words the *slopes* of the magnetic field graph. ('Negative' magnetic field and current levels mean reversing their direction.)

10. 3 The most convenient way to quantify slopes

So far we have only referred to the size of the slope in relative terms: larger or smaller. How might we measure these quantitatively, in other words attach a numerical value to their size?

The 'slope' of a section of a graph means, of course, its '*direction*'. This needs clarification: when the graph is a straight line the direction is clear, but when the graph is a curve, what do we mean by '*the*' direction if the direction keeps changing?

First no slope should be quoted without identifying the point along the curve at which that slope occurs. The point is identified by the value of the free variable. As the slope of a curve is not a fixed quantity it must be a *function* (as we have already determined), a function of the same free variable as that of the curve.

One way to measure the slope at a chosen point would be to greatly magnify the curve so that a short section near the point appears straight enough to measure the slope. A more convenient way is to draw a straight line that touches the curve at the chosen point and measure the slope of the straight line. (An obvious name for this line would have been "toucher", but we have to put up with "tangent".)

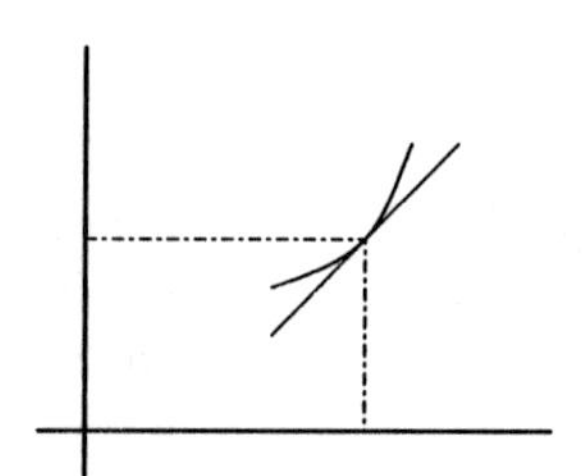

Let us return to how we might quantify the slope with a number. We could simply measure the angle of the slope relative to, for instance, the horizontal axis. This would be somewhat tedious, but possible. The main problem arises when the connection between the variables is given as a *function* rather than as a graph. For example, the connection between height and time for the stone we threw up in the air, which we showed in the form of a graph, could also be stated by a function such as $H = 15\,t - 4.91\,t^2$ (H: height in metres; t: time in seconds). It is difficult to see how we might find out values for *angles* from information contained in the function presented in this form.

There is however something that is *connected* with the angles and that can, quite readily, be evaluated from the function.

As 'angles' are to do with *graphs* but the source of information is a *function*, we focus on the connection between a function and its graph.

We have two problems on our hands: to find the connection between the angles and the function, and to do so for curves where the angle keeps changing. Let us do one thing at a time, and start with an angle ⇔ function of a straight line.

As an example we will look at the straight line (*'linear'*) function

$$Y = 1.6\,X$$

- the vertical axis presents Y, and
- the horizontal axis presents X.

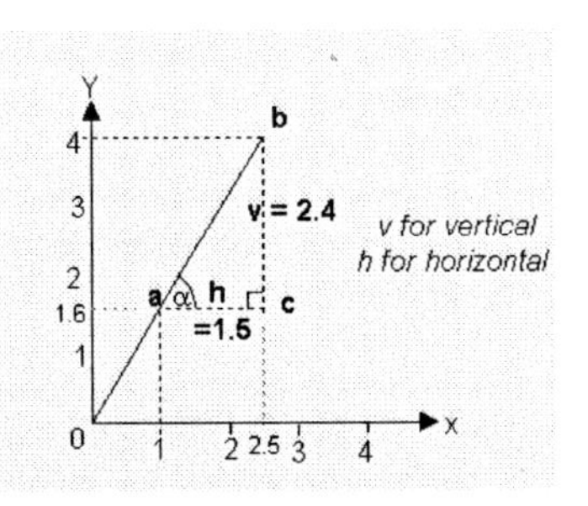

An angle is measured between the directions of two straight lines, in this case between the horizontal direction and the direction of the graph's straight line. A straight line is defined by two points. *Any* pair of points will do because the direction is the same anywhere on a straight line. So, let us choose points 'a' and 'b', which we define as follows: (refer to the diagram!)

'a' is the point whose X value is 1, the Y value is determined by Y= 1.6 X, so Y= 1.6×1 = 1.6

'b' is the point whose X value is 2.5, the Y value is determined by Y= 1.6 X, so Y= 1.6×2.5 = 4

The segment of line whose slope we want starts at 'a' and we evaluate the slope relative to the horizontal, so we draw a horizontal line through 'a'.

We find point 'b' on the graph by choosing its X value, 2.5, and from there draw a vertical line upwards to the meet the graph line.

Looking at all this artwork: the section a-b, the horizontal & vertical lines, we find a 90^0 triangle!

And sure enough, this faithful aid provides us with a connection between angle α and some entities which can be readily obtained from the function:

section v is the OPP of α, section h is the NEXT, and their ratio v/h is the OP/NX ('tan') of α. The 'Short' of this connection is:

OP/NX(α) = v/h. (So, if we know v/h we could find α by searching through the OP/NX table)

Finally, what have h and v to do with the function?

v is the difference between the Y value at point b and the Y value at point a;

h is the difference between the X value at point b and the X value at point a;

Using some 'Shorts':

$$v = Y_b - Y_a$$

$$h = X_b - X_a$$

So,

$$\text{OP/NX}(\alpha) = ({}^{v}/_{h}) = \frac{Y_b - Y_a}{X_b - X_a}$$

Looking at the diagram of our function Y=1.6 X this ratio is $\frac{2.4}{1.5}$ i.e. 1.6; this is the OP/NX ('tan') of 58^0. Find your protractor and check.

But here is the beauty of it all: we do not actually need the diagram.

Remember, each Y is related to its X though Y= 1.6 X, so the $Y_b - Y_a$ (in $\frac{Y_b - Y_a}{X_b - X_a}$)

can be written as $1.6\,X_b - 1.6\,X_a$

which can be compacted with the 1.6 written only once: $1.6(X_b - X_a)$

so the ratio $\frac{Y_b - Y_a}{X_b - X_a}$ (which equals OP/NX(α)) becomes $\frac{1.6\,(X_b - X_a)}{X_b - X_a}$; the equal top & bottom disappear in a puff, leaving behind sheer elegance: 1.6.

So, from the function Y= 1.6 X alone we get the slope angle, having found that its OP/NX is 1.6;

Now, for some generalizations:

Clearly this does not work only with 1.6: if we repeat the above with k in place of the 1.6 the OP/NX of the slope will come out as k.

What about the function Y= k X + d?

Whatever Y came to with Y= k X it is now bigger by an amount d, at every X. This has the effect of 'shifting up' every point on the graph by the same amount, thus generating a line that is parallel to the first one, i.e. has the same direction, that is, the same slope k.

Why did we not use OP/L or NX/L (sin or cos)? Knowledge of their value would reveal the value of the angle just as OP/NX (tan) does. There is a good reason to steer clear of them.

OPP and NEXT can easily be found from (differences between) X and Y values, *precisely what the functions are all about.* Figuratively speaking OP & NX is about 'up & across'; these are the "language" of functions. With OP/L and NX/L however the problem is the L that is involved. Finding its value requires the tedious manipulation of $(OP^2 + NX^2)^{1/2}$ (and then it must be used to divide OP or NX), so why bother if one can get the same service by simply dividing the OP by NX, both directly available from the function.

We established that the general way of getting the slope information from a function is

$$OP/NX(\alpha) = \frac{Y_b - Y_a}{X_b - X_a}$$

We can state the above in a form that shows the connection to the function in a clearer way.

It all results from using two points on the graph. Each point is associated with a value of the free variable X ('free' because we can *choose* its value), and a value of the dependent variable Y ('dependent' because its value is calculated by the function which uses whatever X was chosen). We used the function Y= 1.6X, but in general, if we do not want to restrict ourselves to a *particular* function, we can write Y=*f*(X).

Using this notation Y_a becomes $f(X_a)$ and Y_b becomes $f(X_b)$

and so we re-write as $$OP/NX(\alpha) = \frac{f(X_b) - f(X_a)}{X_b - X_a}$$

We can simplify this further. At present the expression depends on two items (X_a and X_b), but this is not really so: we know that the evaluation of the slope does not depend on where the second item (X_b) is as long as it is somewhere different from X_a. So let us say just that:

$X_b = X_a$ + some (any) increase (or, 'difference') in X.

Employing some 'Shorts': $X_b = X_a + DX$

Thus $\dfrac{f(X_b) - f(X_a)}{X_b - X_a}$ becomes $\dfrac{f(X_a + DX) - f(X_a)}{X_a + DX - X_a}$. Note that the divider simply equals DX.

So: $$OP/NX(\alpha) = \frac{f(X + DX) - f(X)}{DX}$$ (as there is now only one X we can drop the $_a$)

It is useful to describe this in words:

To determine the slope at a point that corresponds to a given value of the free variable X, we evaluate the function (Y) with that X and also with a neighbouring X.

The *OPP*by*NEXT* of the slope angle is equal to the difference between the two values of the function (the latter less the former), divided by the difference between the neighbouring X's.

Or, another possible wording:

A change in the free variable (a horizontal move on the graph) causes a change in the value of the function (a vertical move on graph). The slope's OP/NX is the ratio between these related changes (the latter divided by the former).

Let us check this with our straight line ('linear') function Y= 1.6 X, using a DX 'gap' = 1:

the value of the function at some X is Y = 1.6 X. Now we increase X by 1:

the value of the function at X+1 is Y = 1.6 (X+1), which (written without brackets)

is 1.6 X +1.6. Subtracting from this the value of

the function at the first point X, namely 1.6 X leaves 1.6 (note that the X vanished).

All this is the same as saying that 'Y=1.6 X ' means "when X grows by 1, Y grows by 1.6".

So there is a difference of 1.6 between the Y values of the function at two X points that are 1 apart (i.e. DX=1). Dividing this 1.6 difference of the Y values (call it 'DY') by the 'DX' of 1 gives the already familiar 1.6, the OP/NX of the slope angle (DY acting as OPP, DX as NEXT).

The value of the slope (1.6) contains no X; this means it is the same 1.6 no matter what X is, i.e. no matter where along the graph. Had the slope-expression contained X (thus being "a function of X") it would imply different slope values at different points on the graph. "Same slope everywhere" means "straight line". This is what Y= 1.6 X is, and we noted above how this kind of a function causes the X to 'fall out' during the evaluation of the slope.

If Y=X, i.e. Y=1x X the slope (its OP/NX) is 1. But Y=X also means that a change in X (*h* in our diagram) causes an equal change in Y (*v* in the diagram), making a 2-equisided 90^0 triangle, with $\alpha = 45^0$. And what is OP/NX(45) (tan45) if not 1?

10.4 Direction of a curve?

In the examples given above the graph 'happened' to be a straight line. Generally however graphs are curves.

Drawing tangents can only serve to *measure* slopes on graphs, not to evaluate the slope from the function.

Here is the problem: the slope, as we know, is about the direction of a *straight* line, however short it may be. In order to define a line we need two points. Every point is defined by its X and Y values, the X we can choose freely, but the associated Y comes from the function. What do we do if the function is of a *curve*?

Again a trick is needed. The first of the two points is where we want to know the slope. Then, as the direction of the straight line does not care how short it is, we take the second point on the curve so close to the first one so that the section between them is as good as straight. Admittedly, the slope *between* these two points will not be *exactly* the same as *at* the first point, but it will a) be nearly so, and b) be more and more so the closer we bring the two points together. The great question is: by investigating how the slope is affected by bringing the two points closer and closer, might we by chance find that we could *deduce* what the slope would be if the two points were to unite?

Before we proceed we need to point out the following.

We have equated 'rate of change' with 'slope' because 'rate of change' means:

"increase in the dependent variable *divided by* the corresponding change in the free variable", if instance one unit of it; If a graph is drawn with the dependent variable on the vertical axis and the free variable on the horizontal axis then the definition of 'rate of change' is the same of the OP/NX of the local angle of the graph.

If we actually want to *see* the correct angle (rather than just calculate it), both axes of the graph need to be drawn to the *same scale*.

Consider the two graphs: both are of the same function, with the same rate of change 3 (to 1), but only in the first, where both axes are to the same scale, is the angle of the graph such that its OP/NX (tan) is 3 (72^0)

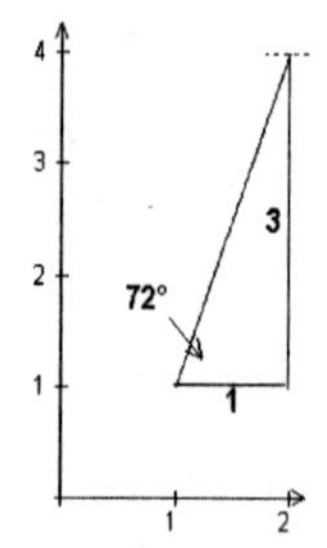

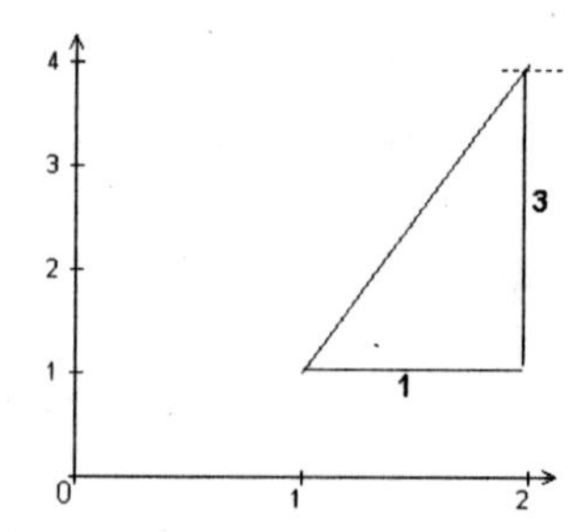

If however we do not want to *measure* anything, but only calculate the OP/NX, that is, the relative change of the dependent and free variables, there is no need for the same scale. Thus if it is more convenient, we can 'spread' one of the axes as in the second graph.

So let us attempt to investigate what happens to the OP/NX when we start by evaluating it from two separate points, and then when we bring the points closer and closer together. We will do this for the function $Y = X^2$.

(When the variables do not represent anything in particular, like time, height etc. it is a convention to use Y for the dependent variable and X for the free variable.)

To specify points on a graph by their X & Y values, we will use the notation X|Y e.g. X=3|Y=9.

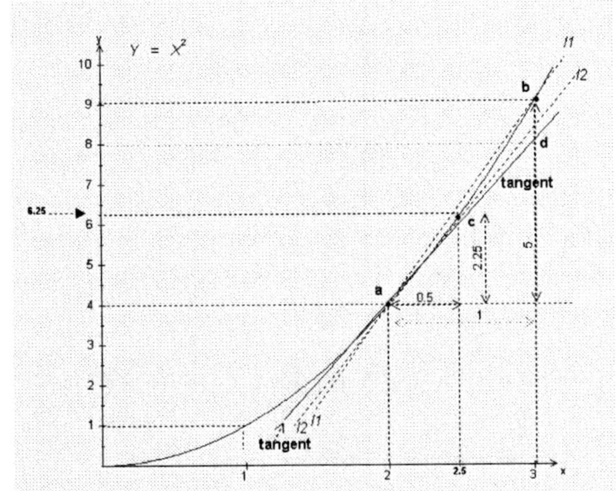

We want to evaluate the slope of the function at X=2|Y=4 ($Y = X^2$)

As a neighbouring point let us begin by choosing X=3|Y=9.

Our purpose here is to work everything out from the function, rather than measure slopes on a graph, the 'picture' is provided only to illustrate what is going on.

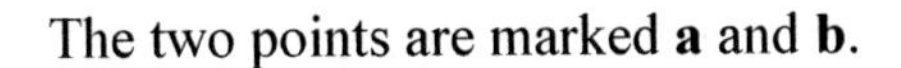

The two points are marked **a** and **b**.

We are concerned with the difference between the Y values of these two points divided by the difference between the X values of the two points.

From here on we will have a lot of this "difference between Y values and between X values", so we had better use the 'D' for short to mean "Difference between two ... values", thus

DY for "difference between two Y values" which results from

DX , the "difference between two X values"

(Note again that if we consider again the 90^0 triangle containing the slope angle α then:

DY is the same as 'OP' and

DX the same as 'NX'.)

We need to fix the 'direction' of these differences, that is, which one less which. Since DX relates to *increase* in X this means the larger X less the smaller one. DY is *not* "larger Y less the smaller one", but "Y *belonging to the larger* X less the Y *belonging to the smaller* X".

We now return to the slope of the function between X=2|Y=4 and X=3|Y=9

so: DY $= 9 - 4 = 5$, and

DX $= 3 - 2 + 1$

and so their ratio DY/DX , which is the same as the OP/NX of the slope angle, equals $5/1 = \mathbf{5}$.

Now, in order to get a better approximation to the slope at X= 2 | Y=4, let us move the second point closer, and place it at X = 2.5 (point **c** in the graph).

Here Y is 2.5^2 which is 6.25.

We are now looking at the slope of the line between **a** (2 | 4) and **c** (2.5 | 6.25):

DY is now 6.25 - 4 = 2.25, and

DX is 2.5 - 2 = 0.5 .

So DY/DX is now 2.25/0.5. We get rid of the point in the divider by multiplying top and bottom by 10, which gives us 22.5/ 5 = 4.5.

In this way evaluation of the OP/NX of the slope has changed from 5 to **4.5**. What is there to stop us getting closer?

Let us now take **X = 2.1** as our position of the second point.

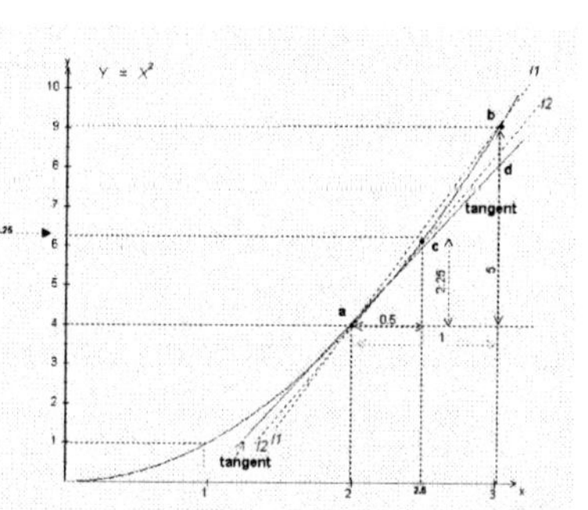

Y here is 2.1^2 = 4.41.

Now the differences of Y and X relative to point **a** (2 | 4) are

DY = 4.41 - 4 = 0.41

DX = 2.1 - 2 = 0.1.

At this point then, DY/DX = 0.41 / 0.1 = **4.1**

We began with slope (OP/NX) 5, then 4.5, now 4.1. Let us look *really* closely at **X = 2.01**.

Its Y is 2.01^2 = 4.0401.

So once again we take the difference from point X=2 | Y=4:

DY = 4.0401 - 4 = 0.0401 and

DX = 2.01 - 2 = 0.01

DY/DX is now 0.0401 / 0.01 (multiplying top and bottom by 100) = 4.01 / 1 = **4.01.**

We begin to *suspect* that if we go on taking the X value of the second point nearer and nearer to **X= 2**, the DY / DX would eventually settle at **4**.

Let us follow this process also in the drawing above:

l_1 is the line joining X=2 | Y=4 to a point **b**: X=3 | Y=9; the slope of this line is 5.

l_2 is the line joining X=2 | Y=4 to a nearer point **c**: X=2.5 | Y=6.25; the slope is 4.5.

't' is the *tangent* drawn at X=2 | Y=4. Tangents are difficult to draw accurately, but this one appears to pass through a point **d** at X=3 | Y=8. In getting there from X=2 | Y=4 the Y rises by 4 while the X grows by 1. Thus here too the slope (OP/NX) appears to be 4.

So it *seems* both from the graph, and, more pertinently, from our calculations, that the slope at X=2 | Y=4 is 4, but we want to *know*.

This brings us to one of the most beautiful things in mathematics.

We are investigating what happens at various distances from X=2, that is, various values of DX. Therefore, rather than use many different *numbers* for DX (as the 1, 0.5, 0.1 above, etc.), let us simply use DX itself as a *variable*.

The neighbouring point to X=2 | Y=4 will thus have an X value of $2 + DX$,

and the Y value there is then $(2 + DX)^2$.

The Y value of the point where the slope is determined we will write, for uniformity, as 2^2 (rather than 4).

The *change* in Y, which we call DY, is then $(2 + DX)^2 - 2^2$.

Therefore the slope $\frac{DY}{DX}$ is then written as $\frac{(2+DX)^2 - 2^2}{DX}$

This can be checked: for instance with the second point at X=2.1 as above, where DX = 0.1,

the slope's DY/DX is $\frac{(2+0.1)^2 - 2^2}{0.1} = 4.1$, as we already found.

Now we can use this expression $\frac{(2+DX)^2 - 2^2}{DX}$ to try and find out what happens at X = 2, that is when DX = 0.

What we then get is $\frac{(2+0)^2 - 2^2}{0}$ which is $\frac{2^2 - 2^2}{0}$ i.e. $\frac{0}{0}$. What is 0/0?

0 divided by anything is 0, and anything divided *by* 0 is ∞ as we know. This means that 0/0 must be either 0 or ∞, or perhaps somewhere in between. "Between 0 and ∞" is one way of saying "*any* number". Not much of a "how much"... , so where do we go from here?

Before we do away with the DX's in $\frac{(2+DX)^2 - 2^2}{DX}$, let us examine what exactly goes on in there, specifically, what the '$(2 + DX)^2$' part of the top actually contains. We know that:

$$(2 + DX)^2 = 2^2 + 2 \times 2 \times DX + DX^2 \quad \text{[i.e. } (a + b)^2 = a^2 + 2ab + b^2, \text{ with a=2 \& b=DX]}$$

and so $(2 + DX)^2 - 2^2 = 2^2 + 2 \times 2 \times DX + DX^2 - 2^2$. (The 2^2 eradicates the -2^2)

leaving $4\,DX + DX^2$

so, our $\frac{(2+DX)^2 - 2^2}{DX}$ becomes $\frac{4\,DX + DX^2}{DX}$. If we divide top and bottom by DX, then on the top the 4DX becomes 4, the DX^2 becomes DX, and the bottom becomes 1.

So, from our $\frac{DY}{DX} = \frac{(2+DX)^2 - 2^2}{DX}$ all that remains is $\frac{4 + DX}{1}$, or, just 4+DX.

Now if we kill off the DX (put it to 0) thus bringing the two points together at X = 2, we do get a clear result:

$$\frac{DY}{DX} = 4 \text{ !!}$$

10.5 The critical bit that usually is not disclosed

Where is the trick in this? Why did we arrive at a meaningless 0/0 above, and now at a very meaningful 4?

By 'unpacking' the $(2 + DX)^2$ into $2^2 + 2\times 2DX + DX^2$ we were able to examine its contents and find out what happens to the ratio $\dfrac{(2+DX)^2 - 2^2}{DX}$ *on the way* as DX goes to 0.

The essence of what is discovered this way is as follows.

While both the top and bottom, that is both $(2 + DX)^2 - 2^2$ and DX, ultimately reach 0 together, the *top always remains greater* (about 4 times greater) than the bottom throughout their joint journey to their grave. Specifically the ratio is, as we saw, 4 +DX, which is 4 when DX = 0.

This can be visualised with the help of the sketch to the right. Both lines meet the ground together, but throughout their way one line is always *relatively* higher than the other, in fact *4 times* higher, all along, including at the end.

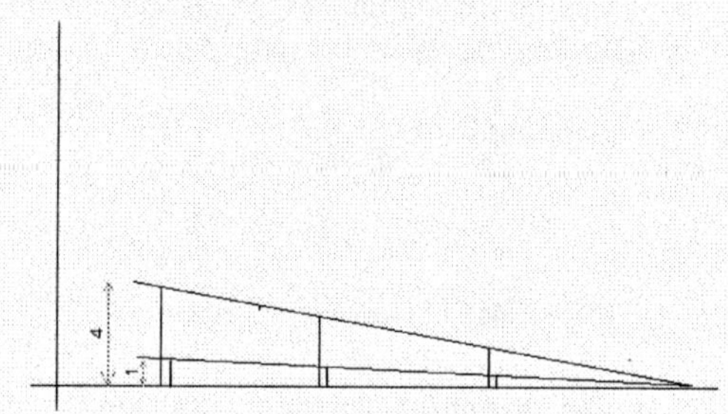

10. 6 Right, you now know enough to do the following

All this was done for the slope at the point corresponding to X = 2. Let us briefly repeat this for slopes at points with *any* X. We will simply do everything just as before, only substituting 'X' for the '2' in $\dfrac{(2+DX)^2 - 2^2}{DX}$

$$\text{so,} \quad \frac{DY}{DX} = \frac{(X+DX)^2 - X^2}{DX}$$

Here the slope is between points

(**a**) $X \mid X^2$ and (**b**) $X+DX \mid (X+DX)^2$.

(remember: the item after the '|' is the corresponding Y value)

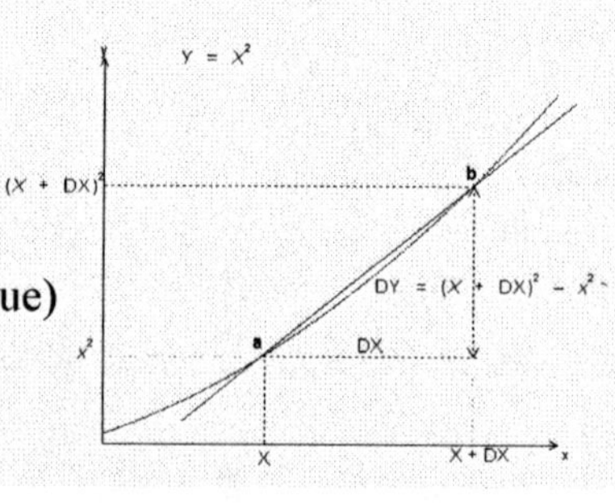

Starting from $\dfrac{DY}{DX} = \dfrac{(X+DX)^2 - X^2}{DX}$:

if as before, we bring the two points together by shrinking DX to 0 we end up with

$$\frac{(X+0)^2 - X^2}{0} \quad \text{which equals the useless 0/0.}$$

So again we open up the brackets *to see exactly what happens in there*, specifically: how fast things move up there relative to the divider when DX is reduced:

$$(X + DX)^2 \quad = X^2 + 2 \times X \times DX + DX^2,$$

so $$(X + DX)^2 - X^2 = X^2 + 2 \times X \times DX + DX^2 - X^2 = 2 \times X \times DX + DX^2$$

so, our $\dfrac{(X + DX)^2 - X^2}{DX}$ becomes $\dfrac{2X \times DX + DX^2}{DX}$.

As before, dividing top and bottom by DX , leaves 2X +DX. Again DX can now die peacefully, leaving a legacy of a clear and healthy

$$\frac{DY}{DX} = 2X\ !$$

What this means is that as X grows so does the slope. For instance: at X is 2, as we found earlier, DY/DX equals **4.**

Other points can be checked on the graph:

At point X=1 | Y=1 the slope (2X with X=1) should be <u>2</u>, as confirmed by the OP/NX of tangent ***a***.

At point X=3 | Y=3^2 (9) the slope (2X with X=3) must be <u>6</u>, as confirmed by the OP/NX of tangent ***b***.

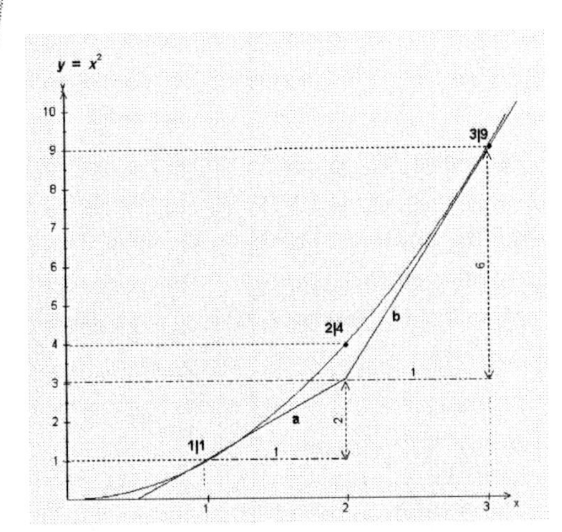

A little note about littleness: Once DX has shrunk the DY is very small too. In tribute to this fact we use the smallest D we know of,

and write: $\dfrac{dY}{dX}$. Remember: the value of $\dfrac{dY}{dX}$ at any point of the function, equals the OP/NX (tan) of the slope angle of the tangent at that point, (provided X and Y are drawn to the same scale).

Note: in '$Y = X^2$' we say that "Y is a function of X". Now, the expression that gives the slope of this function at various X values, we found to be '2X', therefore *this too* is a function of X.

We could call it the 'slope function'; 'they' call it "*derivative*" (lest you might want to know what they are talking about), and the process of finding the derivative is known as "*differentiating the function*", and, to really make you despair before you even start, they call the whole subject "*calculus*".

We now know that the 'slope function' of $Y = X^2$ is $2X$.

What is it for $Y = gX^2$, where g is any number?

Comparing $Y= X^2$ to $Y= gX^2$ the effect of g is to make each Y value g times larger, which is the same as 'stretching' the graph g-fold in the Y direction. For example, take g as 3. Consider (*a*) the graph of *any* function, and (*b*) the graph of the same function multiplied by 3: this would be like taking the paper on which (*a*) was drawn and stretching it 3-fold in the Y direction. Obviously every section on the graph, that is, any section between the same two X values, gets 3 times steeper:

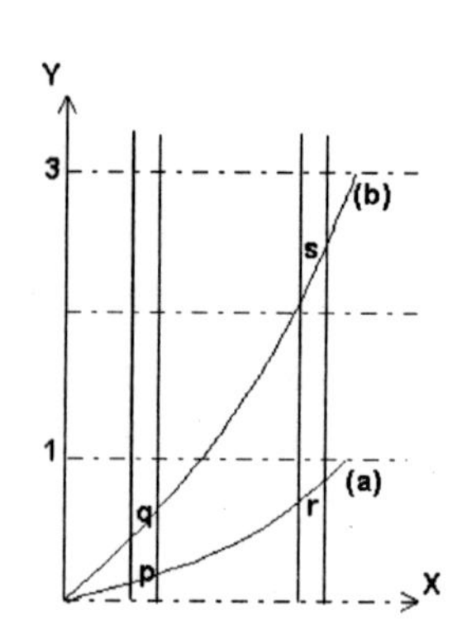

for instance, section q is 3 times steeper that p, and section s is 3 times steeper than r.

Therefore, as we know that for $Y= X^2$ the slope function is $2X$, we can conclude that for

$Y= gX^2$ the slope function is $g \times 2X$ $(2g X)$.

What about $Y= gX^2 + c$?

This, compared to $Y= gX^2$, means raising every point by c, which amounts to 'shifting' the whole curve upwards - by the same amount everywhere, and this of course does not effect the slope anywhere, so the slope function is still $2g X$.

Finally, $Y= gX^2 + kX$.

Unlike adding c which raises Y by the same amount at every X position, 'kX' adds a larger amount the further to the right one goes. Specifically, with every unit advance in the X direction, k units are added to the Y (on *top* of the gX^2). This implies a *slope of k*, which is *added* to the slope of $Y= gX^2$. So,

the slope function (dY/dX) of $Y= gX^2 + kX$ is $2g X$ +**k**.

We could have done all the above in the more 'mathematical' way: writing the Y value at X+DX i.e. $g(X+DX)^2 + k(X+DX)$, opening the ^{2} brackets, then subtracting the Y value at X i.e. the $gX^2 + kX$, then dividing by DX and examining what is left after DX is shrunk to nothing. We would, of course, have arrived at the same result, but doing it the way we did here offers more insight into what is happening. We can make an important generalisation:

$2g X$ was the slope function of $Y= gX^2$,

k, we know, is the slope function of $Y= kX$. We found that

$2g X + k$ is the slope function of $Y= gX^2 +kX$

So, the slope function of a **sum of functions**, is the **sum** of their **slope functions**.

Remember, $2g X +k$ is also the slope function of $Y= gX^2+ kX$ **+c**, the '+c' does not effect the slope.

10.7 Top and bottom, where art thou?

Here is one of the nicest and most useful applications of working out the slope functions.

Let us return, for an example, to the function we met with earlier, giving the height H in metres at various times t, in seconds, for a stone thrown up into the air:

$$H = 15\,t - 4.91\,t^2$$

(15 is the velocity, in metres per second, with which the stone is launched.) Here the dependent variable is H and the free variable t. (H and t are a sensible notation, because the letters recall what these variables are about. However as we are now into Ys & Xs, so, in consistence with the expressions of slope functions we have been studying, we will continue to use Y and X for the present.) Height will be Y and time X. Thus the expression giving the height of the stone as a function of time is $Y = 15\,X - 4.91\,X^2$. For the sake of convenience we will approximate it to $Y = 15X - 5X^2$.

We expect to see the stone again. In other words this function must reach a maximum value at some point, after which the stone begins its descent. But at what point? At what X does a function reach a maximum (if it does so at all)?

Looking at the height graph and, below it, its corresponding slope graph we can see where this happens: where the curve is (momentarily) horizontal. In other words, the slope at this point is 0.

In order to find at what X this occurs we need to establish the slope function of $Y = 15X - 5X^2$ and then work out at what X it equals 0.

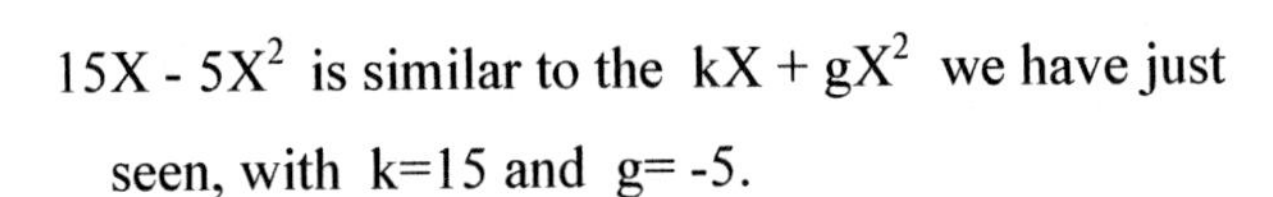

$15X - 5X^2$ is similar to the $kX + gX^2$ we have just seen, with $k=15$ and $g= -5$.

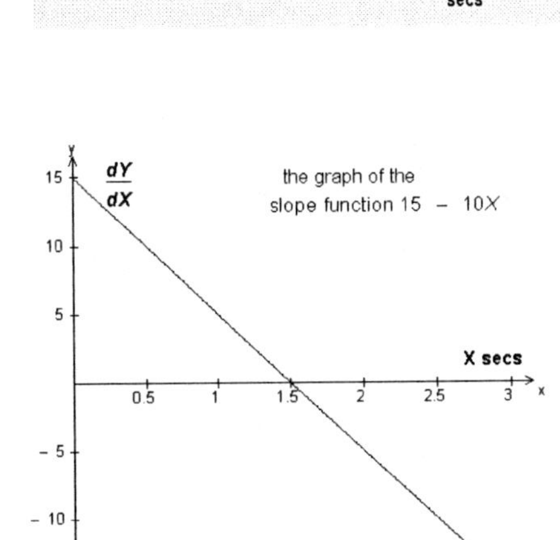

For $Y= kX + gX^2$ we found $dY/dX = k+ 2g\,X$,

so, in the case of

$$Y = 15X - 5X^2 \qquad dY/dX = 15 + 2\times(-5)X$$

$$= 15 - 10X$$

At what X is this $15 - 10\,X$ equal to 0?

This question is best put into shorthand as "find X for $15 - 10X = 0$ ". This gives us an opportunity to solve an equation. Subtract 15 from both sides: $-10X = -15$;

then bare X down by dividing both sides by -10: $X = -15 / -10$ which is 1.5

This is indeed (the X value of) the point at which the graph reaches its maximum.

Now we can also work out *how high* the highest point is. We do this by using - what else - the height function

$Y = 15\mathbf{X} - 5\mathbf{X}^2$

and substituting for X this *1.5* value at which we just found the curve to peak:

$Y_{max} = 15\times\mathbf{1.5} - 5\times\mathbf{1.5}^2 = 11.25.$

This result is confirmed by the graph.

This leaves one question unanswered. A zero slope point need not necessarily be at the top of a curve; it might also be the bottom!

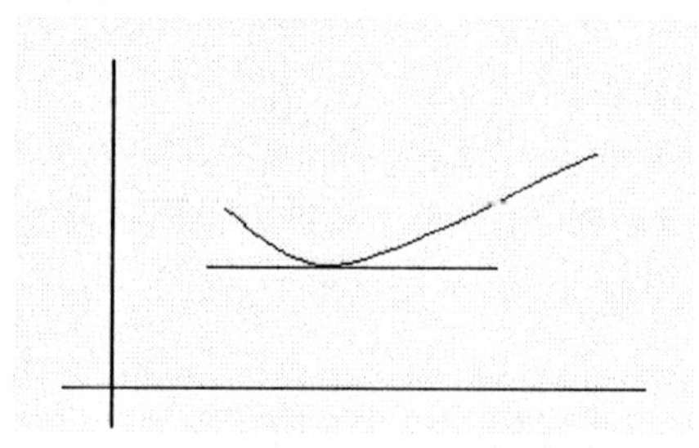

Therefore, when we have a function, and have worked out where the slope is 0, but have not bothered to draw its graph, how can we know whether we have found a peak or a trough?

If the horizontal section of the curve is a **peak** the preceding part of the curve was on the way **up**, and the following part on the way **down**. This, in slope-talk, means that the slope was **positive** just before 0 and **negative** immediately afterwards, that is, the slope graph goes **down**.

What we are talking about here is the ***slope of the slope function***, and here it has a **negative** value, specifically at the X at which the slope function is 0.

If the horizontal section of the curve is a trough every **bold** item in the above story is inverted and so the ***slope of the slope*** function has a *positive* value where the ***slope*** function is 0.

Nothing is to stop us working out the *slope function* of the *slope function*. In the case above the slope function was 15 - 10X. For every X increase of 1 this function 'grows' by -10. This means that the *slope of this slope function* is –10, very negative. In this case this is true everywhere, that is, at any X, and that includes our X = 1.5 (see the graph of the slope function, above). Therefore it is thanks to the **negative slope** of the **slope function** of the **height function** that we can rest assured that the stone eventually drops *down*, not up...

Chapter 11

Maths As Art

Most of what we have encountered so far could be deduced by orderly thinking. Maths can never do without this, but some results are arrived at only with the help of an *idea*, just as in art, and with as much beauty.

This is an opportunity to reveal one such idea in all its glory, an idea that shows imagination and simplicity, and is of great practical importance.

11.1 Fixed increment series

There are two basic kinds of series of numbers. In the first, and simplest, called 'fixed increment series', there is a fixed difference between consecutive numbers in the series. For example in the series 2, 5, 8, 11, 14, 17 the fixed difference (which we will call 'D') is 3. (Its conventional name is "Arithmetic series" - as usual, very insightful as if the other series involve knitting...). Three things are needed to define a series. We need to state where the series starts, how it progresses, and how far it runs, i.e.

the first member	M_1
the fixed difference	D
the number of members	n

Two things must be worked out:

the last member	M_n
the sum of all n members	S_n

(although any two of the above *five* quantities can of course be worked out given the other three.)

All of this is very simple, but the simplest of all is working out the last member, M_n. If the last member is the 5th, there are **4** steps of D leading to it from M_1. And so, if there are n members, it is n-1 of the D that must be added to M_1 to get M_n, that is,

$$M_n = M_1 + (n-1)\,D$$

If we check this in the case of the series 2, 5, 8, 11, 14, 17: M_1=2, n=6, D=3, so

$$M_n = 2 + (6-1) \times 3$$
$$= 17. \quad \text{OK.}$$

As for the sum, S_n, if the whole series were 2, 5, 6, 11 we could just tot them up, but when there are 109 members in the series we would welcome a less laborious way of getting there.

No art is required yet. We can *deduce* the way to go about it.

We are looking for a shortcut for a process of repeated addition. The classic way to do this is multiplication. Multiplication however works best if all the items to be added are the same, which is not quite the case with 2, 5, 8, 11, 14, 17... The natural question then is whether we might find one number which could stand in for each of these six different numbers for the purpose of finding their sum.

Let us call this hypothetical number 'W'; we need a number that would make

$$W + W + W + W + W + W \quad \text{equal to}$$

$$2 + 5 + 8 + 11 + 14 + 17.$$

If such a W could be found our S_n (now S_6) would simply be $6\times W$.

The W has to equally represent the small and the large members of the series, so naturally it must be something in between.

'*Average*' springs to mind. However the method of calculating the average (or 'mean') involves, as we have seen, finding the sum of the members and then dividing it by the number of members. Finding the sum of the members is exactly what we are looking for a method for, so the *sum* cannot be part of the method...

How then *can* we find this common representative W?

The answer, it should be understood, can only emanate from the special *features* of what this is being done for - if we are looking for a special method of summing fixed increment series we can only base it on their special feature, the fixed difference between consecutive members of the series. Any procedure we devise *must* use this fact.

When the whole solution is not immediately obvious one tries to progress by steps. Let us look at just the first and last members of the series, 2 and 17, and find a W to replace the 2 *as well as* the 17, in such a way that the value of $2+17$ will also result from $W+W$. For this, W must be greater than 2 by the same amount that it is smaller than 17, in other words it must be mid-way between 2 and 17, that is, 9.5 (which is 7.5 away from both 2 and 17). *Check*: $9.5+9.5 = 2+17$.

Now let us take the second and the one but last numbers of the series, 5 and 14. Because the differences between 2 and 5, and 14 and 17, are equal, moving up from 2 to 5 gets us nearer to W by the same amount as moving down from 17 to 14, so W is also mid-way between 5 and 14 (4.5 away from each). Thus $W+W$ will also do the same job as $5+14$. For the same reason, W is also mid-way between 8 and 11, thus covering all members of the series. What we have found is that, by moving symmetrically inwards from the extremes, each *pair* of numbers, the small and the large one, can be replaced by the same *pair* of Ws. In other words every *single* member can be replaced by a *single* W (for the purpose of summing).

(This also applies to cases of odd n, where one of the members, the middle one, is itself a W.)

We saw that W started off in such a way that $W+W$ (i.e. 2W) $= 2 + 17$, therefore

$$1W = (2 + 17) / 2 = 9.5.$$

So, the sum of the **6** members, $2+5+8+11+14+17$ equals **6**W, that is, 57.

Naturally, one would not resist the urge to check this sum...

All this can be generalised in the following way:

2 was the first member, M_1, and

17 the last, M_n.

W was (first + last)/2, or, in 'Shorts':

$W = (M_1 + M_n)/2.$

S_n, the sum of n members, each one of which was replaced by W, is

$$n \times W,$$

so, $$S_n = n \times (M_1 + M_n)/2.$$

We must keep in mind that our aim is to express the sum, S_n, in terms of the first member M_1, the fixed increase D, and the number of members n. We are nearly there except that the above expression for S_n 'suffers from' the presence of M_n and the absence of D.

We found earlier how to express the last member, M_n, in terms of M_1, n, *and*, the desired D:

$$M_n = M_1 + (n-1)D$$

All that remains therefore is to replace the M_n by $M_1 + (n-1)D$ in the our expression for S_n

so $$S_n = n \times (M_1 + M_n)/2$$

becomes $$S_n = n \times (M_1 + M_1 + (n-1)D)/2.$$

Collecting the M_1's, and moving the n to the right for convenience: $$= (2M_1 + (n-1)D) \times n/2.$$

If we want, we can open the outer brackets: $$2M_1 \times n/2 + (n-1)D \times n/2,$$

which can be simplified and rearranged to:

$$S_n = M_1 \times n + n(n-1)D/2$$

Again, testing this for 2, 5, 8, 11, 14, 17,
(where $M_1=2$, $n=6$, $D=3$): $S_6 = 2\times6 + 6\times5\times 3/2 = 57.$ OK again.

Now we are ready to tackle a series with 109 members:

$$S_{109} = 2\times109 + 109\times108\times3/2 = 17876.$$

Surely it is nice to get this result by a few multiplications rather than by adding 109 numbers.

And the sum of the first 1000 integers, that is, 1+2+3...+1000 ($M_1 = 1$; **D=1**; $n = 1000$) is

$$S_{1000} = 1\times1000 + 1000\times999\times1/2 = 500500.$$

It is always instructive to check a formula in extreme situations; for instance, a lonely series of one member, starting from 1, i.e. putting $M_1=1$ & $n=1$ in $M_1 \times n + n(n-1)D/2$):

$$S = 1\times1 + 1\times0\times D/2 \quad \text{(= 1, as probably expected).}$$

There is something nice here: due to the '0×' it does not matter what D is, just as one would expect in a case where there is just the first and only member.

11. 2 Fixed ratio series

(conventionally called "Geometric series" because *our* name is too revealing)

In the other type of series, for example 3, 6, 12, 24, 48, 96 it is the *ratio* (R) that is fixed between the members. In this particular series the ratio is 2. M_1 here is 3, and n is 6:

3	6	12	24	48	96
M_1	M_2	M_3	M_4	M_5	M_6

To find the 2nd member, multiply the 1st by R:

$$M_2 = M_1 \times R \qquad (\text{i.e. } 3\times2 = 6)$$

to find the 3rd member, multiply the 2nd by R,
which is the same as multiplying the 1st by R×R:

$$M_3 = M_1 \times R\times R = M_1 \times R^2 \qquad (3\times2^2 = 12)$$

to find the 4th member, multiply the 1st by R again: note : always one less than

$$M_4 = M_1 \times R^3 \qquad (3\times2^3 = 24)$$

etc.,

and the nth member: $M_n = M_1 \times R^{n-1}$

In the example above, the last member $M_6 = 3 \times 2^5 = 3 \times 32 \ (= 96)$.

Note that the expression $M_n = M_1 \times R^{n-1}$ also holds for n = 1 (i.e. where M_n *is* M_1):

$$M_1 \times R^{1-1} = M_1 \times R^0 = M_1 \text{ as expected (since } R^0 = 1)$$

So far so good, but what about the sum?

That is, how can one cut short $S_6 = 3 + 3\times2 + 3\times2^2 + 3\times2^3 + 3\times2^4 + 3\times2^5$, or, to

put it more generally, $S_n = M_1 + M_1\times R^1 + M_1\times R^2 + M_1\times R^3 + \ldots + M_1\times R^{n-1}$?

As a first step we could put M_1 outside a bracket: $M_1(1 + R^1 + R^2 + R^3 \ldots + R^{n-1})$, but then what?

The trick we used in the fixed increment series does not work here: considering the above example the mid-way point

between the 1st and last, that is, between 3 and 96, is 44.5, but

between the 2nd and the one but last, i.e. 6 and 48, it is 27,

and between 12 and 24 it is 18...,

all very different from each other.

Other attempts to find a common substitute are also unsuccessful, so what is to be done?

11. 3 Savour this beautiful idea

This is where the fun begins. What follows can hardly have been deduced; it is the product of someone's brilliant idea.

We will apply it first to n = 5, and later with any n.

We begin by writing down the following series:

$$S_5 \quad = M_1 \quad + \ M_1{\times}R^1 \quad + \ M_1{\times}R^2 \quad + \ M_1{\times}R^3 \quad + \ M_1{\times}R^4 \qquad (1)$$

Below this we write the series with (every term of) both sides multiplied by R:

$$S_5 \times R = M_1 \times R \ + \ M_1{\times}R^1{\times}R \ + \ M_1{\times}R^2{\times}R \ + \ M_1 {\times}R^3{\times}R \ + \ M_1{\times}R^4 \times R$$

Since $R^2 \times R = R^3$ etc. the right hand side becomes

$$M_1{\times}R^1 \ + \ M_1{\times} R^2 \quad + \ M_1{\times} R^3 \quad + \ M_1{\times}R^4 \quad + \ M_1{\times} R^5 \qquad (2)$$

To get a clearer view of what happens next we will copy lines (1) and (2), but this time with the second line shifted one step (the position of one member) to the right:

$$(1) \qquad S_5 \quad = M_1 + M_1{\times}R^1 + M_1{\times}R^2 + M_1{\times}R^3 + M_1{\times}R^4$$

$$S_5 \times R = \qquad M_1{\times}R^1 + M_1{\times}R^2 + M_1{\times}R^3 + M_1{\times}R^4 + M_1{\times}R^5 \qquad (2)$$

Consider the two equations $\quad a = b$
$\quad c = d$

'Subtracting the second equation from the first' means a - c on the left and b - d on the right. As we begin with two equal things (a and b) and from each we subtract equal amounts (c and d) we know that the results must be equal too, that is: a - c = b - d

Looking at our above equations (1) and (2) and the special positioning of their terms, it is immediately obvious that if we subtract one equation from the other, (1) - (2), we will get a startling result: most of the contents on the right just fall away!

On the left we get $S_5 - S_5 \times R^3$

On the right, all that survives is M_1 .... and $- M_1{\times}R^5$

- the four terms in the middle vanish together with our troubles (of having to add them up).

(The surviving $- M_1{\times}R^5$ has a 'minus' because this term comes from the subtracted line).

So, collecting all that is left of the long snakes that were fed to each other:

$$S_5 - S_5 \times R = M_1 - M_1 \times R^5.$$

As our purpose is to find the sum (S_5 =...), we first rewrite the left side with S_5 appearing only once, and, while we are at it, we can also shorten the right, with M_1 too, written only once:

$$S_5 \times (1 - R) = M_1 \times (1 - R^5)$$

Finally we isolate S_5 by dividing (both sides) by (1 - R):

$$S_5 = \frac{M_1 \times (1 - R^5)}{(1 - R)}$$ - the sum of the fixed-ratio series.

Note: 5 - 3 = 2 and 3 - 5 = -2, in other words, reversing the order of subtraction reverses the sign of the result, the same effect as of multiplying by -1. Therefore changing the order of subtraction on the top and bottom is equivalent to multiplying both by -1, which does not change the value of the ratio. Thus we can write the equation above

$$S_5 = \frac{M_1 \times (1 - R^5)}{(1 - R)} \text{ (form A)} \quad \text{also as} \quad S_5 = \frac{M_1 \times (R^5 - 1)}{(R - 1)} \text{ (form B)}$$

What, one may (should) ask, is the point of having both forms?

In the series 3, 6, 12, 24, 48... the R is 2, that is, larger than 1 (as is R^5). Here form B is more convenient, because in form A '(1 - R)' and '(1 - R^5)' would be negative. That is not really a problem, but it is nice to be positive...

R, however, does not *have* to be larger than 1. In the very common series

1, 1/2, 1/4, 1/8, 1/16 etc.

R is ½ ! (M_1 = 1). If R is less than 1 and you multiply it by itself, things get smaller still, thus all the selfmulted R's (i.e. R^n) are also less than 1. In such cases form A provides positive contents in the brackets, for instance (1 - 1/2) and (1 - $1/2^n$) etc..

What then is the sum of 3, 6, 12, 24, 48?

M_1=3, R=2, n=5:

$$S_5 = 3 \times (2^5 - 1) / (2 - 1)$$
$$= 3 \times (32 - 1) / (1)$$
$$= 3 \times 31$$
$$= 93! \quad \text{Nice?}$$

And for the series 1, 1/2, 1/4, 1/8, 1/16 :

M_1=1, R=1/2, n=5:

$$S_5 = \frac{1 \times (1 - \frac{1}{2}^5)}{(1 - \frac{1}{2})}$$

$$= \frac{1 \times (1 - \frac{1}{32})}{\frac{1}{2}},$$

putting 1 and 1/32 on a common divider:

$$= \frac{\frac{32 - 1}{32}}{\frac{1}{2}}$$

$$= \frac{31}{32 \times \frac{1}{2}} = 31/16.$$

Check this, (by putting all the members of the series on a common divider 16).

In conclusion let us establish a generalised expression, that is, one that applies to any n:

the 1st member is M_1,

the 2nd member is $M_1 \times R$, which, for uniformity, will be write $M_1 \times R^1$

the 3rd member is $M_1 \times R^2$

the ith member is $M_1 \times R^{i-1}$ always one less than the member's number

and the last member, the nth, is $M_1 \times R^{n-1}$

The sum then appears as (n-2 is one before the last n-1)

$$S_n = M_1 + M_1 \times R^1 + M_1 \times R^2 + \ldots\ldots + M_1 \times R^{n-2} + M_1 \times R^{n-1} \quad (1)$$

Both sides multiplied by R (same as R^1) :

$$S_n \times R = M_1 \times R + M_1 \times R^1 \times R + M_1 \times R^2 \times R + \ldots\ldots + M_1 \times R^{n-2} \times R + M_1 \times R^{n-1} \times R$$

$$R^{(n-2)+1}$$

$$= M_1 \times R^1 + M_1 \times R^2 + M_1 \times R^3 + \ldots\ldots + M_1 \times R^{n-1} + M_1 \times R^n \quad (2)$$

Now we write (1) again and below it (2), this time shifted one step (position of one member) to the right:

$$S_n = M_1 + M_1 \times R^1 + M_1 \times R^2 + M_1 \times R^3 + \ldots\ldots + M_1 \times R^{n-2} + M_1 \times R^{n-1} \quad (1)$$

$$S_n \times R = \quad M_1 \times R^1 + M_1 \times R^2 + M_1 \times R^3 + \ldots\ldots + M_1 \times R^{n-2} + M_1 \times R^{n-1} + M_1 \times R^n \quad (2)$$

Subtracting (2) from (1) leaves

$$S_n - S_n \times R = M_1 + 0 + 0 + 0 + 0 + 0 + 0 - M_1 \times R^n$$

(coming from the *subtracted* line.)

Gathering the leftovers we get

$$S_n - S_n \times R = M_1 - M_1 \times R^n$$

which compacts to

$$S_n \times (1 - R) = M_1 \times (1 - R^n)$$

isolating the desired S_n :

$$S_n = \frac{M_1 \times (1 - R^n)}{(1 - R)}$$

which can also be written

$$S_n = \frac{M_1 \times (R^n - 1)}{(R - 1)}$$

So, by the totally imaginative way of writing out the series, underneath it writing the series with everything multiplied by the fixed ratio, then 'illuminating' matters by writing this multiplied line *shifted to the right* and subtracting it from the original series, we end up with a neat and simple expression that produces the desired result, something for which no method could be *deduced.*

What will be the end? Will there ever *be* an end?

With the series 1, 1/2, 1/4, 1/8, 1/16... we saw that the sum of the first five members is 31/16. As the series progresses the members become smaller and smaller. If we add a very large number of them, does the sum grow indefinitely, that is, does the sum "diverge", or is there a limit it will never exceed (in which case the sum of the series is said to "converge", and to be 'finite')?

In view of the time needed to try it out with an infinite number of members we will do better to use the *expression* we have for the sum, and see what happens if we insert $n = \infty$.

If we take any number (G) that is less than 1 (i.e. a fraction), and self-multiply it (G^n), this amounts to " a fraction of a fraction of a fraction etc.", so the result becomes smaller and smaller as n grows. Ultimately, with $n=\infty$, that is G^{∞}, there is *nothing* left to speak of.

For example, also $\frac{1}{2}^{\infty} = 0$

This means that for the series 1, 1/2, 1/4, 1/8, 1/16... ($M_1=1$, $R= 1/2$, $n=\infty$):

$$S_{\infty} = \frac{1 \times (1 - \frac{1}{2}^{\infty})}{(1 - \frac{1}{2})}$$

$$= \frac{1 \times (1 - 0)}{(1 - \frac{1}{2})} = \frac{1}{\frac{1}{2}} = 2$$

- the sum approaches 2 but never exceeds it. (This is called the sum asymptotes 2.)

Note: with five members only, the sum is 31/16 (=1.94) - *already* nearly there.

If the ratio R is 0.9, the decreasing series goes 1, 0.9, 0.81, ..., so

$$S_{\infty} = 1 \times (1 - 0.9^{\infty}) / (1 - 0.9) = (1 - 0) / (1 - 0.9) = 1/0.1 = \mathbf{10}.$$

Here is another nice one, we can call it the "ladies' dream":

With R= 0.975 you can go on forever, but the sum will never go beyond thirty-nine something.

But watch out:

What happens to 1 + 1/2 + 1/3 + 1/4 + 1/5 etc. ? (called 'Harmonic series' for a *good* reason, to do with why sensitivity to classical music is innate, not habitual.) Intuitively, this sum of ever decreasing members can not get very far either. In fact, the first *hundred million terms* do not even add up to 19... However, helped by another pretty idea, we are in for a surprise:

We start with 1 and add 1/2. The next *two* terms are 1/3+1/4 which is more than 1/4 +1/4 (=½) so they add (more than) ½.. The further *four* terms are 1/5 +1/6 +1/7 +**1/8** which is more than 1/8 +1/8 +1/8 +1/8, (again = ½), so they too provide (more than) another ½. Next, the *eight,* 1/9 to 1/16, are less than eight 1/16's – another ½, etc. So, each time we double the length of the next sequence we are certain to add (more than) another ½. As we have an endless supply of these 1/n's we can have endless further doubled-length sequences (even though they get astronomically long)- each adding (more than) another ½, so their sum, which is the sum of all 1/n, has *no* limit! But this book *has*:

Here

Index

Further Reading

Once you have read this book, we trust that mathematics is now an open book to you. You are "Demathtified".

To explore mathematics further, or learn more about concepts introduced in this book that you enjoyed, look at QEDs website

www.mathsite.co.uk

Printed in the United States
19692LVS00001B/6